NMR and Macromolecules

ACS SYMPOSIUM SERIES **247**

NMR and Macromolecules

Sequence, Dynamic, and Domain Structure

James C. Randall, EDITOR

Phillips Petroleum Company

Based on a symposium sponsored by
the Division of Organic Coatings
and Plastics Chemistry
at the 185th Meeting
of the American Chemical Society,
Seattle, Washington,
March 20–25, 1983

American Chemical Society, Washington, D.C. 1984

Library of Congress Cataloging in Publication Data

NMR and macromolecules.
  (ACS symposium series, ISSN 0097–6156; 247)

  "Based on a symposium sponsored by the Division of
Organic Coatings and Plastics Chemistry at the 185th
Meeting of the American Chemical Society, Seattle,
Washington, March 20–25, 1983."

  Includes bibliographies and index.

  1. Macromolecules—Analysis—Congresses.
2. Nuclear magnetic resonance spectroscopy—
Congresses.

  I. Randall, James C. II. American Chemical Society.
Division of Organic Coatings and Plastics Chemistry.
III. Title: N.M.R. and macromolecules. IV. Series.

QD380.N57   1984   547.7′046   84–366
ISBN 0–8412–0829–8

ACS Symposium Series

M. Joan Comstock, *Series Editor*

Advisory Board

FOREWORD

The ACS SYMPOSIUM SERIES was founded in 1974 to provide a medium for publishing symposia quickly in book form. The format of the Series parallels that of the continuing ADVANCES IN CHEMISTRY SERIES except that in order to save time the papers are not typeset but are reproduced as they are submitted by the authors in camera-ready form. Papers are reviewed under the supervision of the Editors with the assistance of the Series Advisory Board and are selected to maintain the integrity of the symposia; however, verbatim reproductions of previously published papers are not accepted. Both reviews and reports of research are acceptable since symposia may embrace both types of presentation.

This book is dedicated to Frank A. Bovey,
a pioneer researcher in NMR studies of polymers,
on the occasion of his receipt
of the American Chemical Society award
in Applied Polymer Science,
sponsored by the Phillips Petroleum Company,
March 22, 1983.

FRANK A. BOVEY's research interests center around the application of NMR spectroscopy to the study of structure, dynamics, and morphology of synthetic polymers and biopolymers. He has made major contributions to these fields, primarily through the development of techniques to determine microstructure in homopolymer and copolymer chains and in the discovery and characterization of defect structures in vinyl and related polymers. He is the head of the Polymer Chemistry Research Department at Bell Laboratories, Murray Hill, N.J.

Dr. Bovey was born in Minneapolis in 1918. He received a B.S. degree in chemistry from Harvard in 1940, worked during World War II for the National Synthetic Rubber Corporation, a 3M subsidiary, and entered graduate school at the University of Minnesota as a Rubber Reserve Fellow in 1945. His thesis work, carried out under the direction of I. M. Kolthoff, dealt with the mechanism of free radical polymerization. During this time he worked out the mechanism of oxygen inhibition and discovered oxygen–styrene copolymers.

After receiving his Ph.D. in 1948, Dr. Bovey returned to 3M, now in the Central Research Department, where he was made a research associate in 1955. It was at 3M, during the period from 1956 to 1962, that he conducted his pioneering investigations into the characterization of polymers using NMR techniques. He reported some of the first high-resolution proton NMR spectra of polymers in the late 1950s, at a time when it was still generally assumed that the spectra of macromolecules were too complex to interpret. Exploiting the rich detail in these spectra, he developed NMR techniques that are now used routinely in measuring the microstructure of polymer chains. These NMR methods made possible for the first time the determination and quantification of the stereochemical configurations of noncrystallizable polymers.

In 1962 Dr. Bovey joined Bell Laboratories as a member of the technical staff, and was appointed to his present position in 1967. He continued his detailed studies of polymer structure and conformation at Bell Laboratories, and extended the scope of his work to include investigations of nuclei other than protons, branch analyses in polyethylene, and determination of defect structures in vinyl and related polymers. He continues to have a vigorous research program in the areas of polymer conformations in the solid state, polymer morphology, and the mechanisms of polymer stabilization and degradation.

Dr. Bovey has published more than 125 papers on his exploratory research on polymers and has contributed about 20 chapters to books in the field. He has written or has been coauthor of 10 books, including "Macromolecules, an Introduction to Polymer Science" with F. H. Winslow in 1979 and "Chain Structure and Conformation of Macromolecules" in 1982.

He has served on an ad hoc panel of the National Academy of Sciences for the study of guayule rubber, and is currently serving on the nominating committee for a National Research Council panel on Polymer Science and Engineering. He has recently served on the Chemistry Panel of the National Research Council Fellowship Office; on the National Research Council Evaluation Panel for the Center for Materials Science of the National Bureau of Standards; on the Study Section for the National Institute of General Medicine (NIH) Shared Instrumentation Program; and both on the organizing committee and as a member of the U.S. delegation to the 1979 China–U.S. Bilateral Polymer Symposium (Beijing, October 1979), sponsored by the National Academy of Sciences. He is a member of the award committee for the Baker Award of the National Academy of Sciences. In addition, he serves on the editorial boards of the *Journal of Polymer Science—Polymer Chemistry* and *Polymer Physics Editions*—and is an associate editor of *Macromolecules*.

Many awards and honors attest to Dr. Bovey's contributions to polymer science. He received the Union Carbide Award of the Minnesota Section of the ACS in 1962. In 1969 he received the Witco Award in Polymer Chemistry of the ACS and the Outstanding Achievement Award of the University of Minnesota. In 1974 he was awarded the Ford High Polymer Physics Prize of the American Physical Society. In 1978 he received the Nichols Medal Award of the New York Section of the ACS and delivered the Whitby Memorial Lectures at the University of Akron. His receipt in 1983 of the American Chemical Society Award in Applied Polymer Science, sponsored by Phillips Petroleum Company, was the occasion for the symposium in his honor on which this book is based.

Dr. Bovey is a member of the American Chemical Society, the American Physical Society, the American Society of Biological Chemists, the New York Academy of Sciences, Sigma Xi, and Phi Lambda Upsilon. He was elected to the National Academy of Sciences in 1975.

L. W. JELINSKY
Bell Laboratories
Murray Hill, N.J.

CONTENTS

PREFACE

Recent experimental improvements in NMR spectroscopy have enabled the polymer chemist to determine macromolecular structure more definitively than was considered possible even a few short years ago. The sensitivity and range of NMR techniques are now such that investigations of polymer morphology and dynamic behavior are leading to information parallel to that from NMR solution studies of sequence distributions and configuration. For a number of years, progress in polymer synthesis and determination of physical properties far outpaced developments in establishing the microstructure of macromolecules. Advances in both liquid and solid NMR techniques have so changed this picture that it is now possible to obtain detailed information about the mobilities of specific chain units, domain structures, end groups, branches, run numbers, number average molecular weights, and minor structural aberrations in many synthetic and natural products at a level of 1 unit per 10,000 carbon atoms and below. This proliferation of macromolecular structural information is leading to new insights into the relationships between polymer molecular structure and the solid state structure and, ultimately, to an improved understanding of those molecular structural factors influencing polymer physical properties.

The advent of a combination of quite different NMR techniques including cross polarization, magic angle spinning, improved probes, improved software in conjunction with more efficient computers, and sophisticated pulse sequencings not only has led to high-resolution NMR spectra of solids but also has stimulated the imaginations of those engaged in liquid polymer NMR analyses. An appropriate sequence of pulses in solution studies of polymers leads to spectra where only specific carbon types are observed by selectively nulling those carbon nuclei that have different proton multiplicities. Subsequent spectral editings can lead to subspectra of single, specific carbon types. This technique of obtaining highly specific subspectra from the more complex overall NMR spectrum will have great utility in polymer characterization.

This book presents to the polymer chemist illustrations of the most recent advances in NMR characterization of polymers while at the same time honoring Frank A. Bovey of Bell Laboratories. Dr. Bovey received the 1983 American Chemical Society Award in Applied Polymer Science, which was sponsored by the Phillips Petroleum Company and presented at the ACS National Meeting in Seattle in March 1983. Dr. Bovey is certainly the pioneer of NMR studies of macromolecules and has had a profound influence on the progress and application of NMR in elucidating the molecular structure of macromolecules. In honor of Dr. Bovey, a number of leading investigators in the fields of both liquid and solid state NMR have presented their work herein. Quite different experimental manifestations of the NMR phenomenon, which lead to either polymer sequence, dynamic, or domain structures, are reported.

We hope that "NMR and Macromolecules" will serve as a reference book and guide to the polymer chemist who is interested in polymer characterization. We hope it will serve as well as a tribute to Frank A. Bovey by revealing the extensive and detailed molecular structural information available through a variety of NMR techniques in characterization studies of polymers.

JAMES C. RANDALL
Phillips Petroleum Company
Bartlesville, Oklahoma

OVERVIEW

NMR and Macromolecules

F. A. BOVEY

Bell Laboratories, Murray Hill, NJ 07974

High resolution nmr spectroscopy (^{1}H, ^{13}C, and ^{19}F) of polymers in solution has been employed during the last twenty-five years for the elucidation of the microstructure of their chains. Examples of such studies carried out at Bell Laboratories are reviewed, including the measurement of stereochemical configuration and regioregularity (head-to-tail vs. head-to-head: tail-to-tail isomerism). The use of ^{13}C nmr of epoxidized *trans*-1,4-polybutadiene crystals to establish the morphology of single crystals is described. Nmr is also a powerful means for the observation of chain dynamics; the use of deuterium quadrupolar echo spectroscopy for this purpose is illustrated. Finally, a large and exciting field of polymer nmr studies is that of the structure and dynamics of biomolecules, exemplified by proton nmr investigations of nucleic acid structure and function.

The first studies of polymers (1) were published only about a year after the first reports of the nmr phenomenon in bulk matter by Bloch and Purcell in 1946. The early work dealt with nuclear relaxation and chain dynamics in the solid state (2). In the mid-1950's, when the study of small molecules had reached a fairly advanced state, it was still generally assumed that very large molecules could not give useful spectra even in solution because of their supposedly slow motions, as evidenced by the very high viscosities of their solutions. There was also a feeling that their spectra would be too complex to interpret. In the late 1950's, scattered reports of high resolution spectra of synthetic and biological polymers began to appear (3-6). This trickle became a flood as investigators were able to show that NMR is uniquely powerful in the determination of polymer microstructure, including stereochemical configuration (7,8), geometrical isomerism (9,10), regioregularity (11,12), and monomer sequences in copolymers (6,13,14).

0097–6156/84/0247–0003$06.00/0

Stereochemical Configuration

In Figure 1 is shown the 40 MHz ^{1}H spectrum of two samples of poly(methyl methacrylate), as reported by George Tiers and myself in 1960 (7). Spectrum a is that of a polymer prepared with a free radical initiator; spectrum b is that of a polymer prepared with an anionic initiator, n-butyllithium in toluene. As is now well known, the marked differences in the spectra arise from differences in stereochemical configuration. It is clear from fundamental considerations and from the spectra of small model molecules (e.g. the 2,4-disubstituted pentanes) that the methylene protons of *racemic* (*r*) monomer dyads, i.e. those characterizing a syndiotactic chain (Figure 2, upper left), are equivalent by reason of a two-fold symmetry axis and therefore have the same chemical shift. In the absence of vicinal coupling of main chain protons, as in poly(methyl methacrylate), they appear as a singlet. On the other hand, the methylene protons of the *meso* (*m*) monomer dyads composing an isotactic chain are non-equivalent and have differing chemical shifts; they are expected to appear as an AB quartet. It is evident in Figure 1 that the methylene proton spectrum of the anionically initiated polymer exhibits such a quartet (J_{gem} = −14.9 Hz) centered at 8.14 on the now outmoded τ scale (1.86 ppm from TMS) and that therefore this polymer is predominantly isotactic. In the spectrum of the free radical polymer (a), the methylene proton spectrum is a broad singlet, and so this polymer must be predominantly syndiotactic. (Actually neither polymer is entirely stereoregular and at higher resolution both spectra show additional features from this cause.) Thus, proton nmr is an absolute method and no recourse to x-ray is necessary even if this were possible.

The α-methyl resonances centered at *ca.* 9 τ can be interpreted to give quantitative estimates of isotactic (*mm*) and syndiotactic (*rr*) triad sequences of monomer units and also of the mixed unit *mr*, termed heterotactic, which must occur in chains which are not perfectly stereoregular. In Figure 1, we see three α-methyl proton resonances having the same chemical shifts but very different intensities in each spectrum. They furnish a measure of the triad probabilities.

In the years following these relatively primitive observations, a very large number of vinyl and related polymer systems have been studied by nmr in many laboratories. If the resolving power of the spectrometer is sufficient—that is, if the magnetic field strength is high enough—configurational sequences longer than triad may be observed. With respect to β-methylene groups one may expect to resolve tetrad (Figure 2) and hexad sequences, appearing as a fine structure on the *m* and *r* dyad resonances. One may expect α-groups to be resolved into ten different pentad sequences or possibly as many as 36 heptad sequences. We have the following numbers N(n) of sequences of length n:

n	2	3	4	5	6	7	8	...
N(n)	2	3	6	10	20	36	72	..., or

in general $N(n) = 2^{n-2} + 2^{m-1}$ where m = n/2 if n is even and m = (n−1)/2 if n is odd. Although these longer sequences can be resolved in some proton spectra at

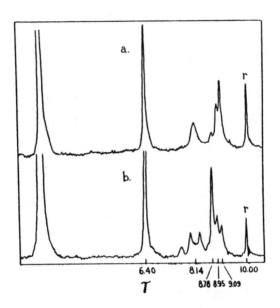

Figure 1. 40 MHz proton spectra of poly(methyl methacrylate) in chloroform; (a) polymer prepared using free radical initiator; (b) polymer prepared using *n*-butyllithium initiator. (Bovey, F. A.; Tiers, G. V. D. J. Polymer Sci., 1960, 44, 173).

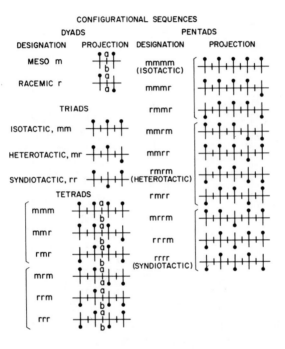

Figure 2.   Configurational sequences in vinyl polymer chains shown in planar zigzag projection.

superconducting frequencies, the use of carbon-13 spectroscopy has proved much more effective, primarily because of the greater chemical shift range of ^{13}C nuclei—over 250 ppm for structures of interest in polymers as compared to less than 10 ppm for protons. Thanks to the development of Fourier transform instruments with spectrum accumulation, carbon spectroscopy, despite the low natural abundance of ^{13}C (1.1%), has become a method of fairly high sensitivity, able to establish the presence of structural features at a level of less than one carbon per 10000.

In Figure 3 are shown the 90 MHz ^{13}C spectra of the CH_3, $\beta-CH_2$ and $\alpha-CH$ carbons of an atactic polypropylene (15). The $^1H-^{13}C$ J-coupling multiplicity has been removed by irradiation of the protons, as is customary in ^{13}C nmr. One may clearly resolve 20 heptad configurational sequences out of a possible 36. Below each spectrum is a line spectrum in which is represented the predicted chemical shift based on the "γ-effect" conformational model (16). In this theoretical model, it is predicated that two carbon atoms separated by three intervening bonds—that is, in a γ position with respect to each other—will shift each others' resonances upfield by about 5 ppm when they are in a *gauche* conformation, as compared to their resonance positions when they are *trans*. Thus, stereochemical configuration influences ^{13}C chemical shifts through its effect on the *gauche* content of the intervening bonds, which may be readily estimated by calculations based on the rotational isomeric state model of the polymer chain.

^{19}F NMR Observations of Regioregularity

^{19}F chemical shifts are also very sensitive to chain microstructure, sometimes even more so than those of ^{13}C. In Figure 4 are shown 84.66 MHz ^{19}F spectra of poly(vinyl fluoride) (17). Spectrum (a) is that of a commercial polymer; the four upfield groups of resonances at 190-200 ppm (from CCl_3F), and some small resonances in the principal spectrum as well, correspond to inverted or head-to-head:tail-to-tail ("syndioregic" (18)) monomer units:

$$-CH_2-CHF-CH_2-CHF-\overset{\displaystyle -}{CHF}-CH_2-CH_2-CHF-CH_2-CHF-$$

Quantitative measurement shows about 11% of the monomer units to be inverted. The principal spectrum shows splitting into *mm*, *mr*, and *rr* triad resonances with some pentad fine structure. The polymer is nearly atactic. Assignment of inversion "defect" resonances is made easier by reference to spectrum (b), which is that of poly(vinyl fluoride) prepared by the following route (17):

$$CH_2{=}CFCl \xrightarrow[\text{initiator}]{\substack{\text{free} \\ \text{radical}}} {+}CH_2{-}CFCl{\rightarrow}_n \xrightarrow{Bu_3SnH} {+}CH_2{-}CHF{\rightarrow}_n$$

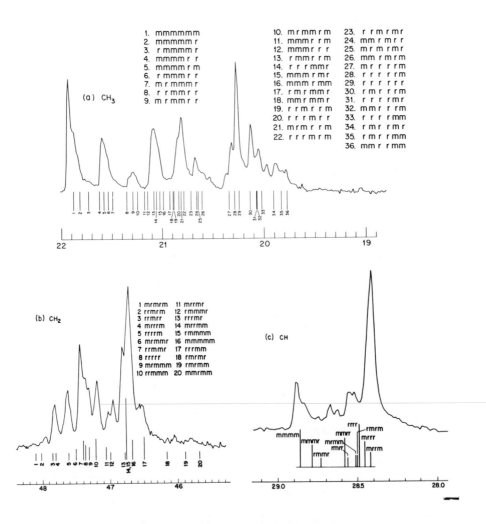

Figure 3.   90.5 MHz ^{13}C spectra of atactic polypropylene observed on a
20% (w/v) solution in heptane at 67°; (a) CH_3; (b) $\beta-CH_2$;
(c) α-CH. Line spectra appearing below each experimental
spectrum correspond to theoretically calculated resonance
positions for (a) heptad, (b) hexad, and (c) pentad
configurational sequences. (Schilling, F. C., private
communication).

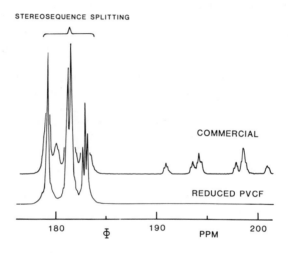

low—field ^{19}F NMR at 84.66 MHz

STEREOSEQUENCE SPLITTING

COMMERCIAL

REDUCED PVCF

180        $\bar{\Phi}$ 190   200
                              PPM

Figure 4.   84.66 MHz ^{19}F spectrum of (a) commercial poly(vinyl fluoride); (b) poly(vinyl fluoride) prepared by reductive dechlorination of poly(vinyl chlorofluoride); both observed at 130° in 8% (w/v) solution in N,N—dimethylformamide (Cais, R. E.; Kometani, J., private communication).

The steric requirements of the chlorine atom permit only a negligible proportion of syndioregic units in poly(vinyl chlorofluoride) (PVCF), and it is now observed that when the chlorine is reduced with tri-*n*-butyltin hydride the resulting poly(vinyl fluoride) exhibits no upfield resonances, being now entirely regioregular. The random configuration evident in spectrum (b), however, does not reflect the stereochemistry of the precursor PVCF but is rather the result of racemization at the α-carbon during the reduction, which is a free radical reaction.

Carbon—13 Study of Crystal Morphology

A quite different use of high resolution nmr in polymer science is the application of ^{13}C spectroscopy to the study of the morphology of chain-folded polymer single crystals grown from dilute solution (19). The problem posed here is the direct measurement of the lengths of the polymer folds at the crystal surface and of the crystal stems within the body of the crystal. The approach is a chemical one, the problem being to find a reagent which will react completely with the exposed folds but will not attack the crystal stems. No such reagent is known for polyethylene, the usual testbed for new morphological approaches, but for crystalline *trans*-1,4-polybutadiene (m.p. 148°) the formation of oxirane rings by reaction with *m*-chloroperbenzoic acid appears to satisfy the requirements of selectivity and convenience:

The crystals, prepared from heptane solution, were suspended in toluene and reacted at 6° until the rate levelled off, which occurred at *ca.* 12-16% of completion. The reacted crystals were then dissolved in CDCl$_3$ and the ^{13}C spectra observed at 50 MHz. The spectra were interpreted with the aid of the spectrum of *trans*-1,4-polybutadiene reacted in homogeneous solution. The statistics of the reaction proved strikingly different in the two cases. The chains of the reacted crystals were in effect block copolymers of quite regular structure with runs of oxirane units alternating with runs of unreacted monomer units, as shown in Figure 5, whereas the reaction in homogeneous solution was random. By comparison of the intensity of the "junction" methylene resonance, D in Figure 5, with that of the internal methylene B (split because these runs are diastereoisomeric sequences of left- and right-handed oxirane rings), it is found that the fold length is only 2.5-3.0 butadiene units. This is the minimum number for a 180° turn of the chain, and appears to require adjacent re-entry. The deduced crystal structure is shown in Figure 6. The stem length can also be obtained by analagous nmr measurements and turns out to be 15 butadiene units for this particular preparation. (The inclination of the stems cannot be deduced from nmr but comes from x-ray).

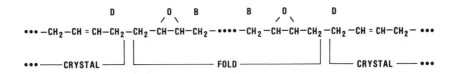

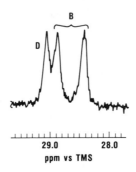

Figure 5.   50.3 MHz ^{13}C spectrum of methylene groups of *trans*-1,4-polybutadiene epoxidized to 16.2% in the crystalline state; observed in CDCl$_3$ solution at 40°. Peak assignments are indicated in reference to the schematic block sequence above, with methylene D representing the junction between sequences. (Schilling, F. C.; Bovey, F. A.; Tseng, S.; Woodward, A. E. Macromolecules, 1983, 16, 808).

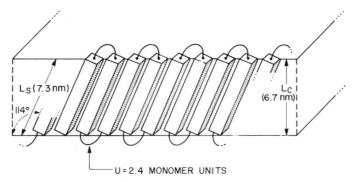

Figure 6.   Schematic diagram of a *trans*-1,4-polybutadiene crystal. The fold length is U and is approximately 3 butadiene units. The stem length L$_s$ corresponds to 15 monomer units. The crystal thickness L$_c$ is obtained from this value and the x-ray determined inclination angle of 114° (Schilling, F. C.; Bovey, F. A.; Tseng, S.; Woodward, A. E. Macromolecules, 1983, 16, 808).

Chain Dynamics; Deuterium Spectoscopy

It is well known that nmr is a powerful means for the study of the dynamics of polymer chains both in solution and in the solid state. The relaxation of ^{13}C nuclei has been extensively employed for this purpose in this and other laboratories. I illustrate here a different and particularly intriguing approach which as yet has seen only very limited application to synthetic polymers. This is *deuterium quadrupolar echo spectroscopy*, as employed in our laboratory by Dr. Lynn Jelinski and her collaborators ([20]). The presence of the nuclear electric quadrupole lifts the degeneracy of the two deuterium Zeeman transitions, and in the solid state produces a very broad (*ca* 200 kHz) powder pattern of transitions which can be interpreted to yield very specific motional information for those carbons labelled with deuterium. In Figure 7 are shown spectra of poly(butylene terephthalate) deuterated on the central carbons of the aliphatic chains:

By comparison of observed and theoretically calculated spectra it can be shown that these carbons are involved in *gauche-trans* conformational jumps of the C-D bond through a dihedral angle of 103°, and from the correlation times as a function of temperature an activation energy of 5.8 kcal/mol is found. Several seemingly plausible motional models are excluded by these results, but the data agree with models proposed by Helfand ([21,22]) for motion about three bonds.

Biopolymers

Finally, a very large and exciting field of polymer nmr studies is that of the structure and dynamics of biomolecules. Here, proton spectroscopy (at superconducting frequencies) remains dominant. I illustrate this by an example of Dinshaw Patel's studies of nucleic acids and oligonucleotides, which shows the use of $^1H-^1H$ nuclear Overhauser enhancement (NOE) to explore the binding site of the antibiotic netropsin to the dodecanucleotide d(CGCGAATTCGCG), which forms the double helical structures

```
C G C G A A T T C G C G
1 2 3 4 5 6 6 5 4 3 2 1
G C G C T T A A G C G C
```

(Here, A stands for adenosine, T for thymine, G for guanosine, and C for cytidine) Because of its sixth-power dependence on interproton distances, the NOE is a very sensitive means of exploring molecular contacts through interproton distances. For large molecules such as these, the enhancement is negative. In Figure 8 are shown difference spectra for the aromatic protons of the netropsin complex with this nucleotide,

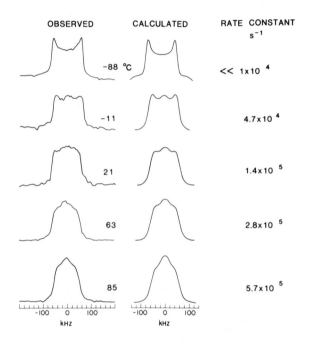

Figure 7.   Solid state deuterium NMR spectra of labelled poly(butylene terephthalate). Calculated spectra are for a two-site hopping model between two orientations of the C-D bond differing by 103° (Jelinski, L. W.; Dumais, J. J.; Engel, A. K. Macromolecules, 1983, 16, 492).

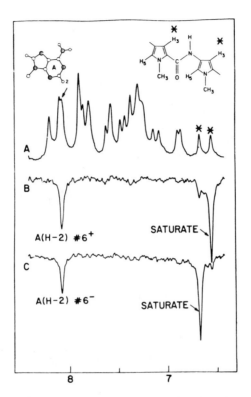

Figure 8.   498 MHz proton NMR spectrum of the 1:1 netropsin-
d(CGCGAATTCGCG) complex in D_2O (pH 6.9, 30°). A
negative NOE at the adenine H-2 proton of the dA·dT base
pair 6 is observed on irradiation of either the 6.55 ppm or the
6.67 netropsin pyrrole H-3 proton. (Patel, D. J.; Pardi, A.;
Itakura, K. <u>Science</u>, 1982, <u>216</u>, 581).

exhibiting large negative peaks for the adenosine H-2 protons of the central position 6 on each strand of the duplex upon saturation of the H-3 protons of each of the two pyrrole rings (they differ) in the bound netropsin (23). These and additional NOE measurements establish that the concave face of the bowshaped netropsin molecule binds in the minor groove of the AATT tetranucleotide core of the complex.

Literature Cited

1. Alpert, N. L. Phys. Rev., 1947, 72, 637.

2. Slichter, W. P. Adv. Poly. Sci., 1958, 1, 35.

3. Saunders, M.; Wishnia, A.; Kirkwood, J. G. J. Am. Chem. Soc., 1957, 79, 3289.

4. Saunders, M.; Wishnia, A. Ann, N.Y. Acad. Sci., 1958, 70 870.

5. Odajima, A. J. Phys. Soc. Jap., 1959, 14, 777.

6. Bovey, F. A.; Tiers, G. V. D.; Filipovich, G. J. Polym. Sci., 1959, 38, 73.

7. Bovey, F. A.; Tiers, G. V. D. J. Polymer Sci., 1960, 44, 173.

8. Nishioka, A.; Watanabe, H.; Abe, K.; Sono, Y., 1960, 48, 241.

9. Chen, H. Y. Anal. Chem., 1962, 34, 1134, 1793.

10. Golub, M. A.; Fuqua, S. A.; Bhacca, N. S. J. Am. Chem. Soc., 1962, 84, 4981.

11. Wilson III, C. W. J. Polym. Sci., 1963, Part A1, 1305.

12. Wilson III, C. W.; Santee, Jr., E. R. J. Polym. Sci., 1965, Part C8, 97.

13. Bovey, F. A. J. Polymer Sci., 1962, 62, 197.

14. Harwood, H. J.; Ritchey, W. M. J. Polym. Sci., 1965, Part B3, 419.

15. Schilling, F. C., private communication.

16. Tonelli, A. E.; Schilling, F. C. Acc. Chem. Res., 1981, 14, 233.

17. Cais, R. E.; Kometani, J., Preprints of 28th IUPAC Macromolecular Symposium, Amherst, Mass. July, 12-16, 1982.

18. Cais, R. E.; Sloane, N. J. A. Polymer, 1983, 24, 179.

19. Schilling, F. C.; Bovey, F. A.; Tseng, S.; Woodward, A. E. Macromolecules, 1982, in press.

20. Jelinski, L. W.; Dumais, J. J.; Engel, A. K. Macromolecules, 1983, 16, 492.

21. Helfand, E.; Wasserman, Z. R.; Weber, T. A. J. Chem. Phys., 1980, 13, 526.

22. Helfand, E.; Wasserman, Z. R.; Weber, T. A.; Runnels, J. H. J. Chem. Phys., 1981, 75, 4441.

23. Patel, D. J.; Pardi, A.; Itakura, K. Science, 1982, 216, 581.

RECEIVED November 10, 1983

NMR OF SOLID POLYMERS

An Introduction to NMR Spectroscopy of Solid Samples

DANIEL J. O'DONNELL

Phillips Petroleum Company, Bartlesville, OK 74004

Complex solid sample NMR techniques used in other chapters of this text are discussed. These techniques are presented using conceptual arguments, rather than mathematical equations, so that those unfamiliar with NMR spectroscopy might get a quick grasp of the nature of the experiment. Figures and charts are given which depict the interactions present in the solid state and which show how these interactions are manipulated in the NMR experiment to yield the desired information.

The papers presented in this volume represent a fraction of the applications developed from new techniques in NMR spectroscopy over the past decade. This flood of new methods has generated new terms which, though mathematically well defined, are difficult to visualize physically. As a result, the advantages of the newer methods (and the information to be gleaned from them) are often lost to the scientist not intimately familiar with NMR spectroscopy.

The objective of this chapter is to provide a guide for the NMR layman to the methods used in the following chapters. Mathematical derivations have been avoided in favor of descriptions and diagrams to provide conceptual definitions. The discussion will concentrate on the techniques used in the following chapters, and therefore will not attempt to be comprehensive. Readers are referred to several texts and articles for detailed treatments of the subjects (1-5).

High Resolution NMR Spectroscopy of Solids

Haeberlin (5) expressed the prevailing attitude of the spectroscopist in 1976 with the statement, "Narrow is beautiful." Although this attitude still prevails, it has long been recognized that a wealth of information is contained in NMR line shapes broadened by specific interactions in solids. Extracting that information has been difficult, since a variety of mechanisms contribute to the line shape, and each must be selectively removed from the others to decipher the information. The methods used to deconvolute the complex line shapes involve manipulations to remove some broadening while retaining other information. Several of these techniques are discussed below.

Contributions to NMR Line Widths in the Solid State

Dipole-Dipole Interactions

Each NMR-active spin $\frac{1}{2}$ nucleus in an external magnetic field, H_O, acts as a magnetic dipole which aligns with H_O in specific states. For protons, carbons and other spin $\frac{1}{2}$ nuclei, two states exist: parallel or anti parallel to H_O. Since each nucleus, as a dipole, has a local field associated with it, the actual field each nucleus experiences is a sum of the external field, H_O, and contributions from all the surrounding dipoles. This dipole-dipole interaction is highly dependent upon the angle between the direction of H_O and the internuclear vector between a dipole pair, and is also highly dependent upon the distance between the dipoles. In an abundant nuclear species such as protons, each proton has many neighbors which interact with it and, because of the dependance of angle and distance, a wide variety of dipole-dipole interactions are possible. In liquids, these specific interactions are averaged by motions which constantly change angles and distances, resulting in narrow lines. In solids, the angles and distances are fixed, resulting in an enormous number of different local magnetic environments, each of which is observed in the NMR spectrum. The differences in local fields induced by dipole-dipole inter-actions result in a range of signals in the NMR spectrum, covering 40 KH_z in some cases. As was mentioned previously, information about internuclear distances can be obtained from the dipole-dipole interactions between two isolated dipoles. The problem is in decoupling all the other dipole-dipole inter-actions, so that one dipole-dipole coupling can be observed.

Chemical Shift Anisotropy

The second major contributor to NMR line widths in spectra of solid materials is chemical shift anisotropy, CSA. CSA results from the interaction between magnetic fields from electrons in motion around a nucleus and the nuclear spin. The distribution of electrons around the nucleus will depend upon chemical bonding, and as a result will not be uniform in all directions. As an example, consider a carbonyl bond. Electronically, this bond is highly directional, and how it interacts with the static field, H_0, will be strongly dependent upon the angle between the bond and the direction of H_0. Rapid motions in the liquid state result in the observation of a net average interaction (i.e., narrow lines). In the solid state all possible orientations are "frozen" in place, resulting in a wide variety of local interactions which are observed in the spectrum (i.e., wide lines). Although information concerning bonding and molecular symmetry are contained in the CSA line shape, the overlap of CSA line shapes in a complex system makes interpretation difficult. The line widths induced by CSA interactions are directly proportional to H_0, so that higher fields do not improve resolution.

Nuclear Quadrupole Effects

The discussion so far has centered on the non-symmetric distribution of local fields surrounding a given nucleus resulting from other nuclei (dipole-dipole interactions) and from electrons (chemical shift anisotropy, CSA). In addition, the nucleus itself may not be symmetric, resulting in a non-symmetric nuclear charge distribution, i.e., an electric quadrupole moment. For nuclei with a spin quantum number, I, of ½ (e.g., ^{1}H and ^{13}C) the quadrupole moment is zero, and no consideration need be given to quadrupolar interactions. However, in the case of deuterium, I equals 1, and the effects of the quadrupole must be taken into account. Since the quadrupole is electric in nature, it interacts <u>directly</u> only with electric field gradients and <u>not</u> with the magnetic field. However, the <u>magnetic</u> energy levels of the nucleus are coupled to the quadrupolar energy levels, resulting in splittings in the NMR spectra. These splittings can be very large, resulting in spectra ~200 KH_z wide for ^{2}D, and 5 MH_z for ^{14}N. At this point one might ask why quadrupolar interactions should be considered, since they are difficult to observe, and in the case of deuterium, require isotopic labelling. The answer lies in the fact that if the electric field gradient about the quadrupole is fluctuating due to molecular motion, line narrowing occurs which is extremely sensitive to the frequency and amplitude of that motion. The reader is referred to the chapter by Jelinski and co-workers in this volume, and references therein for examples and further discussions.

Line Narrowing Methods

In the above discussions of the source of line-broadening in solid state NMR spectroscopy, it was pointed out that all of the interactions were highly orientationally dependent. In the case of dipole-dipole interactions, the orientation of the internuclear vector between two dipoles with respect to H_o gave rise to wide lines. For CSA interactions, the orientation of the chemical bond with respect to H_o was important. Finally, the orientation of the nuclear electric quadrapole with respect to the surrounding electric field gradient gave rise to split- tings in the NMR spectra. In every case it was noted that molecular motion, especially random rotational and translation motion as experienced in the liquid state, resulted in line narrowing by time-averaging. To obtain narrow line-widths in the solid state, it is then up to the spectroscopist to mimic the effect of motion in a liquid to time-average the different interactions.

Dipolar Decoupling

The simplest means of removing dipolar interactions between two nuclei is to decouple the interaction by a means entirely analogous to scalar decoupling used in ^{13}C NMR spectroscopy to remove J-coupling from bonded protons(6). A strong radio frequency (rf) pulse at the resonance frequency of the protons is turned on during acquisition of the carbon signal. The decoupling rf pulse promotes rapid spin transitions or flips between spin states by the proton spins, thereby averaging the static dipolar interactions to zero. This decoupling constitutes a rapid random motion in "spin space", as opposed to the random motion characteristic of molecules tumbling in real space. Unfortunately, it can only be used to remove heteronuclear dipole-dipole interactions. In cases where the isotopic concentration of NMR active species is low, such as ^{13}C (1.1% of all carbons), homonuclear dipole-dipole inter- actions are not significant. The line-shapes left in the ^{13}C NMR spectra of solid when dipolar decoupling of the protons is used therefore normally only reflect the chemical shift anisotropy (CSA).

The Magic Angle

Although dipolar decoupling removes dipolar interactions, it does not remove CSA, nor does it permit observation of abundant spin species such as protons, since observation and decoupling cannot be done at the same time. Different, more selective means are needed to remove these interactions.

It can be shown (1-5) that the magnitude of any of the above anisotropic interactions have a very specific angular dependance with respect to: 1. the static field (CSA), 2. other nuclear spins (dipole-dipole interactions) or 3. with surrounding electric field gradients (quadrupole interactions). Among other terms describing the orientational dependence of these anisotropic interactions in each of the respective Hamiltonian operators is the term $(3\cos^2\Theta -1)$. The angle Θ has a different meaning depending upon the type of interaction being considered. For dipole-dipole interactions, the angle is between a vector joining two dipoles and the direction of H_0 (Fig. 1A). In the case of CSA interactions, the angle might be between the bonding axis and H_0 (Fig. 1b). Finally, the angle in quadrupolar interactions is between the quadrupole moment and the direction of the electric field gradient (Fig. 1c). In each case, if the angle Θ is chosen such that:

$$3 \cos^2\Theta -1 = 0$$

all anisotropic contributions to the NMR spectrum will reduce to zero. This angle, 54.7^o, is justly called the magic angle. Since all values of Θ are possible in the non-crystalline or powdered solid, very few interactions naturally reduce to zero. However, if, over time, the average value of $\Theta = 54.7^o$, the anisotropic static contributions will again reduce to zero. This can be done in real space by high speed sample rotation at an angle 54.7^o from the direction of the field(3); or in spin-space, by manipulation of the spins using rf pulses(7). In Fig. 2, a representation of the effect of spinning on the time averaged value of an internuclear vector is shown. Rotation about the axis R causes the nuclei and the internuclear vector to circle the axis. Over a period of one rotation, the average position of each nuclei lies along R, and the time average internuclear vector will therefore also be aligned with R. The anisotropic static interactions mentioned above will be coherently modulated at spinning rates less than ~ one half of the line width (in Hz), resulting in spinning side bands. At spinning rates greater than the line width, the side bands disappear. Theoretically, it is possible to remove all of the above anisotropic interactions by sample spinning at the magic angle. Realistically, spinning rates have been limited by material problems, so that rates of 3 to 5 KH_z are normal. These rates are sufficient to attenuate or remove line-broadening due to CSA, but are far short of the ~20 KH_z necessary to remove dipolar interactions, with a few exceptions(1).

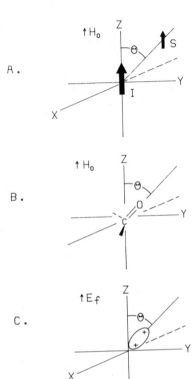

Figure 1. Orientational dependence of anisotropic interactions on the angle θ: a. Dipole-dipole interaction; b. chemical shift anisotropic interaction; c. electric quadrupolar interaction.

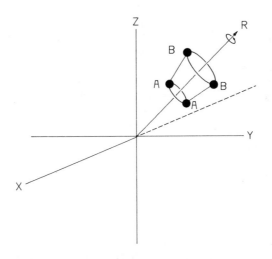

Figure 2.   The time-averaged result of rotation of a randomly oriented internuclear vector, ab, about a rotational axis, R.

Line Narrowing in Spin Space - WAHUHA

Sample spinning induces a time dependent effect on the relationship between a nuclear spin fixed along H_0 and its
"surroundings" by moving the "surroundings" about the magic
angle. If the "surroundings" are now fixed, the same type of
time dependent behavior can be induced by specific rotation of
the nuclear spins about the magic angle. In Fig. 3, a representation of magic angle sample spinning is given from the
viewpoint of the sample. Since the magic angle vector is a
locus of points equidistant from all three coordinates, a view
down the magic angle reveals the three coordinate axis to be
equally spaced about the "magic" vector. As the sample spins,
the applied field, H_0, appears to be "rotating" through each of
the coordinates. The nuclear spins remain aligned along H_0, so
that the spin magnetization, M, also appears to be rotating
through the three coordinates.

In 1968, Waugh and co-workers(7) devised a means of "rotating"
M through each of three coordinate axis. Because nuclear spins
precess about H_0 at the Larmor or resonance frequency, the coordinate system used when spin manipulations are investigated
must also rotate at the Larmor frequency. This also permits us
to view rf pulses at the Larmor frequency as applied fields
along the axes of this rotating frame (when the rotating frame
is used, the axes will be designated x', y' and z').

Clasically, when an rf pulse is applied, e.g. along the x'
axis, this rf field, H_1, will cause the net spin magnetization,
M, aligned along H_0 to precess as shown in Fig. 4a. A pulse of
specific duration for a given H_1 will cause M to precess
exactly ¼ revolution, so that M lies on the y' axis (Fig. 4b).
Changing the phase of the of pulse $180°$ will cause M to precess
in the opposite direction, returning it to the z' axis (Fig.
4c). The same type of pulse sequence can be applied along the
y' axis, rotating M into the x' axis. By proper use of pulses
and delays, M can be made to spend equal amounts of time on
each of the rotating frame axes. As can be seen, this mimics
in spin space the results obtained by sample rotation in real
space. The necessary pulses are diagramed in Fig. 4d. If the
pulses are short enough and the delays, τ, are kept to a minimum, the "rotation rate" of the process can be made fast
enough to effectively decouple homonuclear dipole-dipole interactions. This technique and other, more complicated pulse
sequences(1) have been used to narrow lines in proton spectra.
The method is also important in that it removes only
homonuclear dipole-dipole broadening, but does not effect
heteronuclear dipole-dipole interactions. In the chapter by
Schaefer and coworkers in this text, the WAHUHA pulse sequence

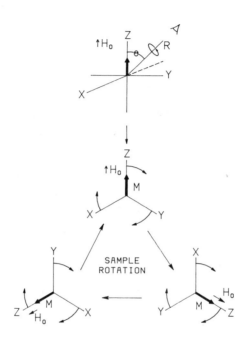

Figure 3.   View down the rotational axis, R, of the net
magnetization, M, and the applied field, Ho.   The viewer
is rotating ccw with respect to the coordinate system.
The angle of R from all three axis is 54.7°.

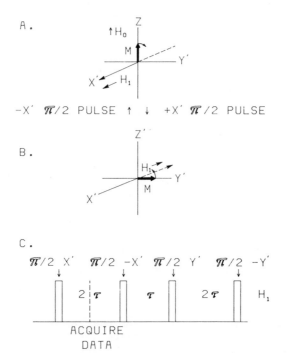

Figure 4. The WAHUHA experiment: a. The action of a
π/2 rf pulse applied to a spin system along the +x' axis
in the rotating frame; b. the action of a π/2 pulse
applied along -x' axis. c. The WAHUHA pulse sequence.

was used to remove $^1H-^1H$ dipole couplings, while retaining $^1H-$
^{13}C couplings.

Finally, it should be pointed out that the above discussions
are vast over-simplifications which are (hopefully)
conceptually easy to grasp. Unfortunately, they are not
theoretically satisfying. Additional reading is urged to provide a more indepth understanding(1-5).

Cross Polarization

Classically, the net magnetization, M, of a spin system results
from the sum of the individual spins which precess about the
applied field, H_O (see Fig. 5a). The frequency of the
precession, called the Larmor frequency (ω_L, Fig. 5a), is
dependent upon the magnetic moment of the individual spins, and
the amplitude of H_O. In addition, the amplitude or polarization of M will depend upon the value of the nuclear magnetic
moment as well. In the case of a proton spin system, the
nearly 100% natural abundance and large magnetic moment result
in a polarization that gives a large net magnetization, M_H
(Fig. 5a). The magnetic moment of a ^{13}C spin, on the other
hand, is about one fourth as large as the proton moment,
resulting in a net magnetization, M_c, that is one fourth that
of protons in the same H_O for an equal number of nuclei. To
make matters worse, the natural abundance of ^{13}C nuclei is only
1.1%. Finally, the length of time for the carbon spin system
to recover from the perturbation necessary to make a measurement can be 10 to 100 times longer than that time needed for a
proton spin system in the same molecule (see spin-lattice
relaxation, below). This creates a time bottleneck when
repeated samplings are taken of the ^{13}C magnetization.

From the above discussion, it is evident that if the proton
spin system could be used as a source of magnetization and as a
means of relaxation for the carbon spin system, an enhancement
of the carbon NMR signal and a savings in time could be
achieved. This energy transfer in the static field, H_O, is not
possible due to the large mismatch in the Larmor frequencies
for the two different nuclei. In order for a transfer to
occur, the two spin systems must have some precessional components that are equal in frequency. As long as the spins are
precessing about the static field, H_O, this is impossible.
However, specific magnetic fields may be applied by using radio
frequency (rf) pulses. In Fig. 5a, the magnetization of the
protons was shown in a coordinate system set in the laboratory;
the laboratory reference frame. An rf pulse at a given
frequency in the laboratory frame will appear as a field of
magnitude H, rotating about the Z axis at the applied

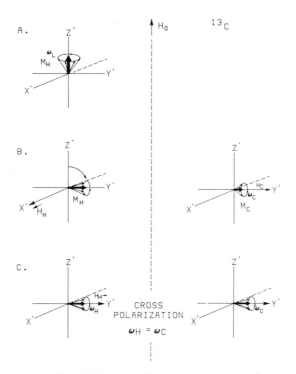

Figure 5.   Spin locking of the proton spin system and
subsequent   cross   polarization   between   the   proton   and
carbon spins: a.   The initial proton magnetization on
the z' axis, aligned with the field, Ho; b.   A $\pi/2$ pulse
is applied along the x' axis of the proton rotating
frame, followed by an rf field, H_C, applied along the y
axis of the carbon reference frame; c.   Spin locking of
the proton spins by H_H and cross polarization with the
carbon spins.

frequency. If the coordinate system were rotating at the applied frequency (a rotating reference frame), then H_1 would appear as a field aligned along either the x' or y' axis. In Fig. 5b, an rf field at the proton frequency, H_H, is shown along the x' axis. This field, as shown in fig. 5b, applies a torque to M_H, the net proton magnetization, making it rotate into the y' axis. At that point, the rf field can be shifted $90°$, so that it also lies along the y' axis (Fig. 5c). Because H_H is a magnetic field that is co-linear with the net magnetization, M_H, the individual spins represented by M_H will precess about H_H at a <u>new</u> precessional frequency, ω_H, that is determined by the amplitude of H_H. If another rf field is applied at the <u>carbon</u> frequency, such that the precesional frequency, ω_C, of the carbon spins about H_C equals ω_H, then an efficient transfer of magnetization can occur (Fig. 5c).

The process of holding a net magnetization along an axis of the rotating frame using an rf field is called spin-locking. The match of rf field intensities H_H and H_C such that $\omega_H = \omega_C$ is called a Hartman-Hahn match(8). The process of transferring magnetization from the proton spins to the carbon spins is called cross polarization(9). The rate of transfer is characterized by a contact time, T_{cp}, which Schaefer and coworkers used to study spin-spin contribution to $T_{1\rho}$ (see below and chapter by Schaefer, <u>et al</u>, this volume).

The cross polarization, or CP, process may be used with any or all of the line narrowing techniques to obtain NMR spectra of solids with resolution approaching that of liquids(3). A combination of cross polarization (CP), magic angle spinning (MAS) and dipolar decoupling were used to obtain the spectrum of a very insoluble polyphenylene sulfide (Ryton) as shown in Fig. 6.

NMR Measurements of Motion in Polymers

All of the interactions discussed above have a strong
dependency on motion, (or lack of motion) in solids. It is not
surprising then that the motions that are present in a solid
can be detected by motion-sensitive NMR methods. The methods
used in this volume can be divided into two catagories; line-
shape analysis and relaxation studies.

Line-Shape Analysis

Because of the strong orientational dependancy of the line-
broadening mechanisms discussed above, the NMR line shape will
reflect changes in the orientation with time. If the frequency
of a motion in a solid is lower than the range of frequencies
(i.e., line-width in H_z) induced by an interaction (e.g.,
electric quadrupolar interactions), a small, but measurable
narrowing might occur. As the frequency and amplitude of the
motion is increased by heating the sample, additional line
narrowing will occur. Dramatic line-shape changes occur as the
motional frequency nears the line-width frequency. Above the
line-width frequency, substantially narrower line shapes are
observed. Conversely, the molecular motion may be rapid compared to the line width at room temperature (narrow lines) and
the temperature dependency of the line shape may be best investigated by lowering the temperature. In this text, Dr.
Jelinski and coworkers used line-broadening from nuclear electric quadrupolar interactions to investigate motions about a
bond between carbons isotopically enriched with deuterium.
This study was particularly interesting, because two regions of
the polymer existed; one that gave a relatively narrow NMR
line-width ("fast" motion) and one that yielded a much broader
NMR line width ("slow" motion). This indicated two regions in
the polymer with distinctly different motional characteristics.

Schaefer and coworkers, in another chapter in this text, used
^1H - ^{13}C dipole-dipole "line shapes" obtained in a very clever
way to investigate rotational motion of the aromatic rings in
polystyrene. The method used a WAHUHA pulse sequence to
decouple proton-proton dipolar interactions, cross polarization
to enhance signal acquisition and an overall sampling technique
synchronous with the sample rotation. The ^{13}C - ^1H dipole-
dipole interaction was mapped in rotational sideband spectra
obtained from 16 "normal" CP/MAS spectra. The method, though
somewhat involved, provided a measure of dipole-dipole line-
shapes which can be interpreted in terms of side-chain rotation
in the polymer.

Relaxation and Motion in Solids
Spin-Lattice Relaxation - T_1

In Fig. 5a, a representation of the net magnetization, M, of a
nuclear spin system is given. The value of M is the sum of the
individual spins precessing at the Larmor frequency, ω_L, about
the applied field, H_0. The individual spins represent an
excess of spins in the low energy state of a system where the
spins are distributed (Boltzman distribution) between two
states (spin $\frac{1}{2}$ nucleus). The system, as shown in Fig. 5a, is
at equilibrium. If an rf pulse, H_1, at ω_L is now applied, the
system will absorb some of the energy from the pulse (the
resonance condition), that is, the system will "heat up". In
Fig. 5b this is represented by a tipping of the magnetization
in the rotating frame. The magnitude of M along H_0 is equal to
zero in Fig. 7b. The system is clearly not in equilibrium with
H_0, and if the rf field, H_1, is removed at this point, the
system will be left in a disordered state. To re-establish the
equilibrium condition, some of the individual spins must
exchange energy, or "cool down." The probability of a spin
giving up energy in the form of discrete rf radiation (a
phonon) is very low. Thus, since energy must be conserved, the
energy must be dissipated in some other form. It may be
dissipated as thermal energy to the atomic framework, or
lattice, if a suitable mechanism exists for the transfer. The
mechanism must be magnetic in nature, and must fluctuate at the
Larmor frequency (i.e., it must fluctuate at a radio frequency
in the megahertz range). The magnetic field sources available
on the atomic scale are nuclear dipoles and unpaired electrons.
If there is rotation, vibration or translation in the lattice
at the Larmor frequency, ω_L, then relaxation of the spin system
to the equilibrium state can occur by passage of the excess
energy to the lattice system; i.e., spin-lattice relaxation
occurs. The relaxation process usually appears to be
exponential in nature, and is usually characterized by a spin-
lattice relaxation time, T_1. The value for T_1 represents the
time necessary for the magnetization to return to within (1-
(1/e)) or 63% of its magnitude at equilibrium. Thus, for full
restoration of M along the field, H_0, one must wait several
times the value of T_1 (usually $5 \cdot T_1$ is sufficient).

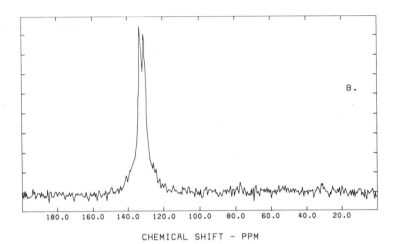

CHEMICAL SHIFT - PPM

Figure 6.   A CP/MAS NMR spectrum of Ryton (polyphenylene
sulfide).   The signals occur at ~135 and ~133 ppm (from
an external reference of TMS).

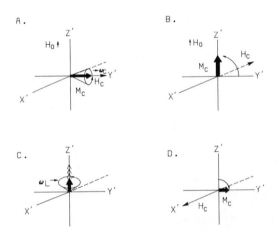

Figure 7.   A mapping of the carbon magnetization during a
T_1 measurement: a.   The carbon magnetization, M_C, locked
on the y' axis by an rf field, H_C, after cross polarization; b. M_C aligned along the applied field, Ho, after a
$\pi/2$ rf pulse along -x';   c. M_C after T_1 relaxation for a
time period, τ;   d. M_C, after a $\pi/2$ rf pulse, aligned
along y'.   This permits the amplitude of M_C to be
measured.

For the individual interested in molecular motion, the
important feature of spin-lattice relaxation (or other
relaxation mechanisms) is the dependency on molecular motion to
provide an efficient energy pathway for relaxation. Thus,
molecular motions at the Larmor frequency for individual carbon
atoms in a molecular framework may be mapped by T_1 measurements. Since the frequency of molecular motion is temperature
dependent, additional thermodynamic and kinetic information may
be obtained by measuring T_1 values for different carbons over a
range of temperatures. In the paper by Lyerla and coworkers in
this volume, T_1 measurements made for the first time over a
range of low temperatures yielded specific information about
motion in the backbone and side chains of a semi-crystalline
and a glassy polymer. The data was taken at a Larmor frequency
of 15.1 MHz for ^{13}C nuclei. It was also noted in this paper
that the T_1 values measured for the glassy polymer showed nonexponential behavior. As stated in the paper, this represented
a distribution of T_1 values due to the many different environments, and therefore the many different T_1 mechanisms, present
in the polymer.

Relaxation in the Rotating Frame-$T_{1\rho}$

As valuable as T_1 measurements can be in analyzing molecular
motions, they suffer from a significant drawback; to map
motions at other frequencies, different magnetic field
strengths, H_0, must be used. This requires the use of a
different magnet (i.e., a different spectrometer operating at a
different Larmor frequency). In addition, the limitation of
the T_1 measurements to molecular motions in the 10^7 hertz frequency range (for ^{13}C) restricts the use of T_1 studies to only
a few types of motions. Motions in the kilohertz region are
therefore inaccessible.

The frequency range of motion available for study by T_1 mechanisms is governed by the precessional (Larmor) frequency of
the individual spins about the applied field, H_0. If a means
could be found to make these spins precess about a much lower
field in the presence of H_0, then the frequency dependent
nature of the relaxation mechanism could be altered to allow
motions in the KHz range to be studied. Such a means has
already been discussed -- spin locking (see Fig. 5). In the CP
experiment, the proton spins are first "locked" along an rf
field, H_H, rotating at the Larmor frequency. As shown in Fig.
5b, H_H appears as a magnetic field aligned along an axis of a

coordinate system rotating at the proton Larmor frequency <u>about</u> H_o (the ^1H rotating frame). As shown in 5c, the individual porton spins will precess about H_H at a frequency, ω_H, <u>directly proportional to the amplitude of H_H</u>. In addition, the carbon spins at the end of the cross polarization transfer are also "locked" along an rf field, H_C, such that the individual carbon spins are precessing about H_C at a frequency, ω_C, <u>proportional to H_C</u>. There are two important features about the spin-lock condition that make it attractive for relaxation studies;

1. The locking rf fields can have a range of amplitudes. Thus a range of motional frequencies can be investigated by simply changing the amplitude of H_H or H_C.

2. The magnetization locked along the applied rf fields are at a magnitude generated by a much larger field, H_o. This means that the magnetization, M, is much too large in proportion to the rf fields, H_H and H_C, and M must diminish via a relaxation mechanism to a level that matches the amplitude of H_H or H_C.

The process of relaxation of M to a value proportional to the applied rf fields in the rotating frame is called <u>spin-lattice relaxation in the rotating frame</u>. The mechanisms available for this form of relaxation are entirely analogous to those available for simple spin-lattice relaxation as described above. Similarly, rotating frame relaxation is characterized by a time constant analogous to T_1, and is called $T_{1\rho}$, or the spin-lattice relaxation time in the rotating frame. Typically, $T_{1\rho}$ values obtained from protons in solid samples are not of much use, since communication between the abundant protons tends to average the relaxation process, so that individual proton relaxation mechanisms cannot be observed. For ^{13}C, however, the natural low abundance of ^{13}C limits the degree of communication, and separate $T_{1\rho}$ values can be obtained for each observed carbon species.

As noted in the papers by Schaefer and coworkers and by Lyerla and coworkers, $T_{1\rho}$ data may be complicated by the fact that mechanisms other than spin-lattice interactions (namely spin-spin relaxation) are possible which don't map motional characteristics. In the case of polystyrene, Schaefer and coworkers conclude that the spin-spin contributions are negligible, whereas Lyerla and coworkers find that the levels of spin-spin and spin-lattice contributions for isotatic polypropylene and atatic polymethyl methacrylate were temperature dependent.

Measurement of Relaxation

The methods to measure ^{13}C T_1 and $T_{1\rho}$ in the solid state are somewhat unique. The initial step for both experiments consists of enhancement of the ^{13}C magnetization via $^{13}C - ^1H$ cross polarization (Fig. 5). At the end of the cross polarization period, the carbon magnetization is either allowed to continue to interact with the rf field, H_C while the proton rf field is removed (Fig. 7a), or an rf pulse is applied to rotate it to align with H_O (Fig. 7b). In the first case, the magnetization aligned along H_C decays via a $T_{1\rho}$ mechanism, and the rate of decay is monitored by changing the length of time H_C is applied. The pulse sequence used is outlined in the paper by Schaefer and co-workers and is repeated here (Fig. 8c). The reader is referred to other references for a more detailed account(3).

If the magnetization M_C is moved to the z' axis to align with H_O, the individual spins will now precess at ω_L around H_O (MH$_z$ frequencies). Remembering that the magnetization represented by M_C is the result of a four-fold enhancement from the cross polarization with the proton spins, then it is evident that M_C does not represent the normal equilibrium magnetization for the carbon spin system. Relaxation occurs (Fig. 7c), and sampling of the process is done by rotating the remaining magnetization via a $\pi/2$ rf pulse into the x', y' plane (Fig. 7d).

The pulse sequences for measuring T_1 and $T_{1\rho}$, including the CP enhancement pulses, are given in Fig. 8.

Conclusions

This short overview of solid sample NMR techniques has been an attempt to explain the experiments used in this text in terms the layman can understand. As can be seen by the titles of the papers on solid sample NMR spectroscopy, the main thrust of research in this area as applied to polymers is the investigation of motions in polymers. The experiments outlined in this and following chapters have been shown to be useful for investigating motions covering a frequency range of 10^4 to 10^8 Hz. Most of these experiments are pre-programmed into the commercial instruments available today. New and even more exciting experiments are possible with the flexibility afforded by the state of the art computer controlled systems. Hopefully, as these instruments become more commonplace, these

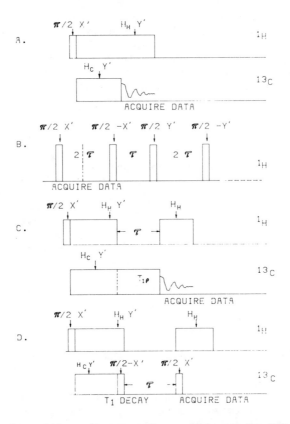

Figure 8.   Pulse diagrams for solid sample NMR experiments: a. ^{1}H-^{13}C cross polarization with dipolar
decoupling; b.   WAHUHA ^{1}H - ^{1}H dipolar decoupling; c.
$T_{1\rho}$ relaxation sequence (time period, τ, is varied); d.
T_1 relaxation sequence (time period, t, is varied).

types of experiments will gain the wider acceptance and use
they deserve.

Literature Cited

1. Mehring, M. "High Resolution NMR Spectroscopy in Solids";
    Diehl, P.; Fluck, E.; Kosfeld, R., Eds.; NMR, Vol. XI,
    Springer-Verlag, New York, 1976.

2. Abragam, A. "The Principles of Nuclear Magnetism";
    Clarendon Press: Oxford University, London, 1961.

3. Schaefer, J; Stejskal, E. O. in "topics in Carbon 13 NMR
    Spectroscopy"; Levy, G., Ed.; Wiley-Interscience: New
    York, 1978; Chap. 4.

4. Goldman, M. "Spin Temperature and Nuclear Magnetic
    Resonance in Solids"; Clarendon Press: Oxford University,
    London, 1970.

5. Haeberlin, U. in "Advances in Magnetic Resonance"; Waugh,
    J. S., Ed.; Acedemic: New York, 1976; Vol. I, pg. v.

6. Block, F. Phys. Rev. 1958, III, 841.

7. Waugh, J. S.; Huber, L. M.; Haeberlen, U.   Phys. Rev.
    Letters 1968, 20, 180.

8. Hartmann, S. R.; Hahn, E. L. Phys. Rev. 1962, 128, 2042.

9. Pines, A.; Gibby, M. G.; Waugh, J. S.   J. Chem. Phys.
    1973, 59, 569.

RECEIVED November 18, 1983

Molecular Motion in Glassy Polystyrenes

JACOB SCHAEFER, M. D. SEFCIK, E. O. STEJSKAL, and R. A. MCKAY—Monsanto Company, Physical Sciences Center, St. Louis, MO 63167

W. T. DIXON[1]—Department of Chemistry, Washington University, St. Louis, MO 63130

R. E. CAIS—Bell Telephone Laboratories, Murray Hill, NJ 07974

The amplitudes of ring- and main-chain motions of a variety of polystyrenes have been established from the ^{13}C NMR magic-angle spinning sideband patterns of dipolar and chemical shift tensors. The frequencies of the same motions have been determined by $T_1(C)$ and $T_1\rho(C)$ experiments. The most prevalent motion in these polymers is restricted phenyl rotation with an average total displacement of about 40°. Both the amplitude and frequency of this motion vary from one substituted polystyrene to another, and from site to site within the same polystyrene. A simple theory correlates the observed ring dipolar patterns with $<T_1\rho(C)>$'s.

Rotating-Frame Carbon Spin-Lattice Relaxation

Both spin-lattice (motional) and spin-spin processes contribute to $T_1\rho(C)$. Experimental cross-polarization transfer rates from protons in the local dipolar field to carbons in an applied rf field can be used to determine the relative contributions quantitatively. This measurement also requires a determination of the proton local field. Methods for making both measurements have been developed in the last few years [1,2]. For polystyrenes, the spin-lattice contribution to $T_1\rho(C)$'s is by far the larger. This means that the $T_1\rho(C)$'s can be interpreted in terms of rotational motions in the low-to-mid-kHz frequency range.

[1]Current address: Mallinckrodt Institute of Radiology, Washington University Medical School, St. Louis, MO 63110.

Dipolar Rotational Spin-Echo Experiment

The $T_1\rho(C)$'s of substituted polystyrenes show alterations in
motion due to the substituent (cf, below) but do not show whether
the alterations involve changes in the amplitude or the frequency
of the motion. This distinction can be made by a separate experiment which measures CH dipolar coupling. The strength of a
static dipolar interaction between an isolated ^{13}C and ^1H spin
pair is known if the internuclear distance is known. This is
the usual situation for a directly bonded CH fragment in a solid
polymer. The reduction in the strength of the CH dipolar interaction by molecular motion (of frequency comparable to or
greater than the dipolar interaction itself) therefore becomes a
measure of the amplitude of the motion. In performing such a
measurement on a real system, the condition that the ^1H-^{13}C spin
pair be isolated from many-body proton dipolar coupling is
achieved by phase-shifted multiple-pulse (WAHUHA) ^1H-^1H decoupling
[3]. The time evolution of the carbon magnetization is then
detected under the influence of ^1H-^{13}C coupling alone [4,5].
The resulting carbon signal can be observed with magic-angle
spinning for high resolution [6,7].
   The pulse sequence for this experiment is shown in Figure
1 [8]. The evolution of the carbon magnetization due to
chemical shift effects is refocused after two rotor periods by
a carbon 180° pulse applied after the first rotor period. Under
high-speed spinning conditions, this removes the effect of the
chemical shift tensor. The ^1H-^{13}C dipolar evolution time is
varied with the number of WAHUHA pulse sequences. The spinning
speed is chosen so that an integral number of WAHUHA cycles
exactly fits into one rotor period. In our experiments, this
number was sixteen. (Each WAHUHA cycle took 33 μsec, with
3-μsec 100° pulses, so that sample spinning was at 1894 Hz.
Matched spin-lock transfers were performed at 60 kHz.)

Typical Spectra in the Chemical Shift and Dipolar Dimensions

Some chemical-shift spectra as a function of WAHUHA irradiation for poly(o-chlorostyrene) are shown in Figure 2.
Protonated carbon magnetizations rapidly dephase under as little
as two cycles of WAHUHA irradiation, but are refocused after
sixteen cycles (one rotor period). Magic-angle spinning should
refocus dipolar coupling just as it does chemical shift aniso-

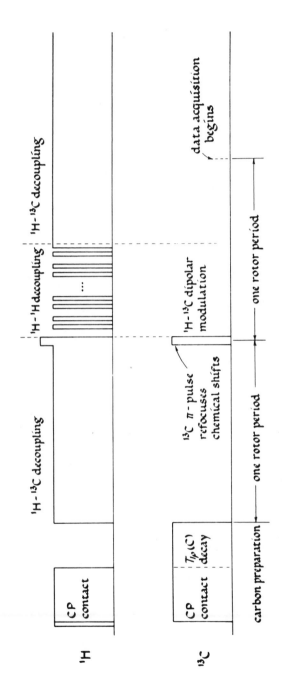

Figure 1. Pulse sequence for a dipolar spin-echo experiment.

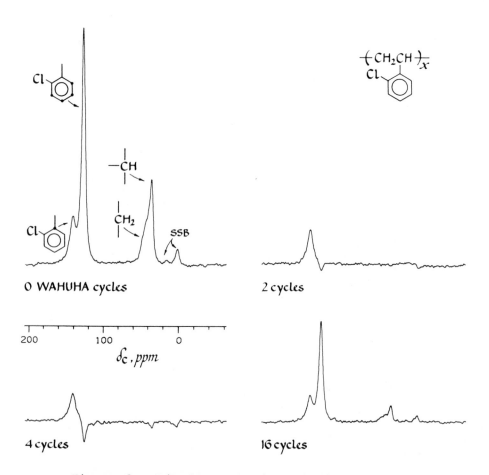

Figure 2.   Dipolar rotational spin-echo 15.1-MHz ^{13}C nmr spectra of poly(o-chlorostyrene) as a function of the number of WAHUHA cycles used during dipolar evolution.

tropy, so in principle, an echo refocused following one rotor period of dipolar modulation should have the same intensity as one without any dipolar modulation. In fact, the former echo is only a third to a half as large. The losses are due primarily to incomplete ^1H-^1H decoupling.

Using 16 separate chemical-shift spectra (the number of WAHUHA cycles varying from zero through fifteen), we performed 16-point Fourier transforms using the WAHUHA irradiation period as the time variable and peak heights as the intensity variables. The resulting transforms for the protonated aromatic carbons of polystyrene, and of crystalline dimethoxybenzene, are shown in Figure 3. To a rough approximation, each spectrum is a Pake doublet [4,5] broken into spinning sidebands [6,7,9] separated by 1.894 kHz. Each sideband is represented by a single point in the frequency domain. Intensities of the sidebands can be reliably compared so long as sideband shapes are the same.

Comparison Between Experimental and Calculated Dipolar Sideband Patterns

The dipolar pattern for polystyrene shown in Figure 3 (left) is not a Pake doublet reflecting only the CH static dipolar coupling (as scaled by WAHUHA irradiation of the protons). Rather, the pattern has been narrowed slightly by molecular motion. Thus, the polystyrene pattern has more intensity in the zeroth and first sidebands then that of dimethoxybenzene, and less in the second sideband. Using the methods developed by Herzfeld and Berger [10], we calculated how ring motion of various types (all restricted ring rotations at a rate fast compared to the dipolar coupling) affected the dipolar sideband patterns. Restricted rotation of an aromatic CH vector about the ring C_2 axis (motion lying on a 60° cone, Figure 4) produces a reduction in intensity of the second, and increases in the zeroth and first dipolar sidebands, with respect to the static pattern (Table I, rows 3 and 4). For poly(p-isopropylstyrene), motion of this type with an rms rotation angle of about 20° produces a reasonable fit with experiment (Table I, rows 2 and 3). In particular, the calculated ratio of intensities of the second to first dipolar sidebands is in agreement with experiment. This ratio is an important parameter to match since it involves the relative intensities of the largest dipolar sidebands, and so is free from normalization uncertainties arising from errors in measuring the weak outer sidebands. The choice of motional model is not crucial for these sorts of small-angle excursions. Motion of the CH vector on a circle rather than a 60° cone gives virtually the same result.

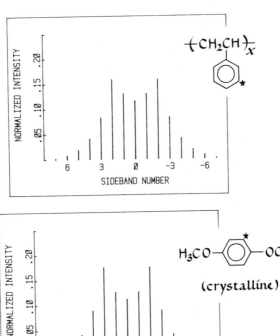

Figure 3.   Experimental dipolar sideband patterns for the protonated aromatic carbon of polystyrene (left) and of crystalline dimethoxybenzene (right) under magic-angle spinning at 1.894 kHz.

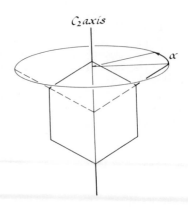

Figure 4.   Representation of the rotation of an aromatic CH vector about the ring C_2 symmetry axis through the azimuthal angle α.

Table I	Comparisons of Calculated and Experimental Dipolar Rotational Sideband Intensities for Poly(p-isopropylstyrene) with Magic Angle Spinning at 1894 Hz					
	sideband number					
experiment or motional model	0	1	2	3	4	5
observed	.138	.142	.153	.079	.038	.014
observed, short-$T_1(C)$ component removed	.129	.135	.156	.083	.041	.015
calculated, C_2 rolls, Θ(rms)=21°	.134	.153	.178	.063	.028	.009
calculated, static	.120	.124	.185	.075	.038	.013

Note than the experimental sideband intensities for poly-(p-isopropylstyrene) have been corrected for the contribution from a few sites at which the rings are undergoing MHz-rate 180° flips. The concentration of these sites for various poly-styrenes can be determined by a $T_1(C)$ experiment, and varies from 0 to 11% for some twelve polystyrenes examined to date [11]. Confirmation that the rings are flipping has been made by studies of on- and off-C_2 axis carbon chemical shift tensors. For our analysis here, it is possible to use the $T_1(C)$ relaxation behavior as part of the carbon preparation step of the pulse sequence of Figure 1, so that dipolar sideband patterns are obtained which are free from contributions from the large-amplitude high-frequency sites in the polymer. Or, knowing the concentration of flipping rings, the appropriately weighted calculated dipolar pattern for a ring undergoing 180° flips can be substracted from the observed pattern for the entire sample, and the difference

dipolar spectrum renormalized. Only such corrected patterns
will be discussed in the following.

Table II	Calculated Ratio of Second to First Dipolar Rotational Sideband Intensities for a CH Pair Undergoing Molecular Motion and Magic Angle Spinning (1894 Hz)	
motional model	total azimuthal angular displacement (deg)	n_2/n_1
	0	1.50
	20	1.44
	40	1.35
	60	1.16
restricted ring rotation[a]	80	0.93
	100	0.70
	120	0.53
	140	0.41
ring π flips	180	0.48

[a] aromatic CH vector undergoing fast diffusional reorientation on the surface of a 60° cone.

Since the second dipolar spinning sideband is near the
cusps of the CH Pake doublet, the ratio of intensities of the
second to first sidebands is a sensitive monitor of partial
collapse of the dipolar tensor by restricted molecular motion.
With a choice of scaling factor to give a static ratio of n_2/n_1
sidebands of 1.50 [8,11], the effect of C_2 axis rotation of
increasing amplitude is shown in Table II. A total angular
excursion of 120° reduces n_2/n_1 by a factor of 3, about the
reduction achieved by 180° flips. The dependence of the n_2/n_1
ratio on the total azimuthal angular diplacement is roughly
linear for displacements greater than about 20°.

Small-Amplitude Low-Frequency Ring Motion

Even after the removal of those sites engaged in ring flipping,
measurable ring motion persists in polystyrenes. In general, as
shown in Table III, the shorter the $<T_1\rho(C)>$, the smaller the
value of $[n_2/n_1]_0^*$. (The star indicates the experimental n_2/n_1
ratio has been corrected to remove contributions from rings
undergoing MHz-rate flips as described in the previous section.
The frequency of flipping is so high there is no significant
contribution to $T_1\rho(C)$.) The fact that both $T_1\rho(C)$ and $[n_2/n_1]_0^*$
show a comparable influence of low-frequency motion more than
hints at a simple connection between them. That is, we can be

polymer	37-kHz $<T_1\rho(C)>^a$, msec	$[n_2/n_1]_0{}^b$	$[n_2/n_1]_0^{*c}$	θ	$\sin^2\theta \times <T_1\rho(C)>$
Table III					

polymer	37-kHz $<T_1\rho(C)>^a$, msec	$[n_2/n_1]_0{}^b$	$[n_2/n_1]_0^{*c}$	θ	$\sin^2\theta \times <T_1\rho(C)>$
poly(p-t-butylstyrene)	6.2	1.04	1.11	23	0.89
poly(p-isopropylstyrene)	8.8	1.08	1.15	21	1.06
polystyrene^d	11.2	1.20	1.25	18	1.00
poly(p-methylstyrene)	11.7	1.20	1.25	18	1.04
poly(α-methylstyrene)	19.0	1.30	1.30	16	1.35
poly(styrene-co-sulfone)	22.0	1.32^e	1.32	15	1.38
poly(o-chlorostyrene)	37.0	1.34	1.34	14	2.02

Table III   Protonated Aromatic $<T_1\rho(C)>$'s of some Polystyrenes Scaled by the Amplitude of Root-Mean-Square Angular Fluctuations Deduced from Experimental Dipolar Rotational Sideband Intensities

[a] straight-line fit to observed decay between 0.05 and 1.00 msec after the turn off of $H_1(H)$.
[b] ratio of intensities of second to first dipolar rotational sidebands with zero $T_1\rho(C)$ decay (Figure 1).
[c] ratio of intensities of second to first dipolar rotational sidebands with contributions from rings undergoing MHz-rate flips removed.
[d] atactic, quenched high-molecular weight material.
[e] contribution to n_1 from the non-protonated aromatic carbon removed assuming $n_1/n_0 = .13$ for that carbon, the same ratio as is observed for polystyrene.

reasonably sure that the same cooperative kHz-regime motions responsible for $T_1\rho(C)$, must also be responsible for the partial averaging of the aromatic CH dipolar tensor.

Correlation Between $\langle T_1\rho(C)\rangle$ and n_2/n_1

For the protonated aromatic carbons of polystyrenes under rotational reorientation, we propose $T_1\rho(C)=K^2\sin^2(\theta)J(\omega)$, where K^2 is a constant (which includes powder averaging in the solid), $\sin^2(\theta)$ is the average dipolar fluctuation orthogonal to the applied rf field (Figure 5), and $J(\omega)$ describes the spectral density associated with the ring motion at the carbon rotating-frame Larmor frequency [12], in this instance, 37 kHz. If we assume ring rotation only occurs about the ring C_2 axis, and if we also assume ring $J(\omega)$'s for all polystyrenes are the same, then the relative $T_1\rho(C)$'s should be a simple function of the amplitude of the ring motion. These amplitudes can be estimated from the reduction in the dipolar CH patterns as characterized by the $[n_2/n_1]_0^*$ ratios, if we assume that the motion which reduces $[n_2/n_1]_0^*$ is also responsible for the $T_1\rho$ relaxation. The results of such a comparison for seven substituted polystyrenes are shown in Table III. The product of $\sin^2(\theta)$ and $\langle T_1\rho(C)\rangle$ is indeed roughly constant for all seven polymers. The product for the first six polymers in Table III is constant to within about 50% even though the $\langle T_1\rho(C)\rangle$'s themselves vary by a factor of 4.

Conclusions for Ring Rotations in Polystyrenes

The ring rotations generate total angular displacements of about 40° (for substituents in the ortho position) to 70° (for bulky non-polar substituents in the para position). Constraints on this motion probably involve both intra-chain steric interactions (for an o-chloro substituent) and inter-chain packing (for polystyrene itself). We suspect the frequencies of many of the small-amplitude ring rotations are determined by fluctuations arising from alterations in local interchain packing. Since the packing changes when the main chains move, the frequencies of small-amplitude ring rotations and main-chain reorientations are comparable. Thus, ring substituents are important in determining the amplitudes but not the frequencies of these types of cooperative ring rotations. This is consistent with our assumption that the $J(\omega)$'s for all polystyrenes are the same.

We have chosen motion about the C_2 axis to model ring motion in polystyrenes. We acknowledge that these motions are likely to be more complicated than just C_2 rotations. However, the motions are small amplitude. Small-amplitude wiggling can equally well be modeled by aromatic CH motion on a circle or sphere, or by C_2 rotations. Thus, any of these

for relative comparisons of polystyrene aromatic $\langle T_{1\rho}(C)\rangle$'s

$$\frac{1}{T_{1\rho}(C)} = K^2 \sin^2\theta \, J(\omega)$$

θ measures fluctuation amplitude due to molecular motion
K includes powder averaging

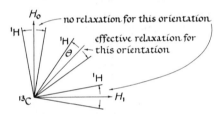

MAS reduces orientational dispersion of rates

Figure 5. Geometrical considerations for a CH vector giving rise to $T_{1\rho}(C)$ relaxation in a solid by restricted rotational reorientation.

models can be considered to represent approximately rotational
excursions by a ring as it tracks its attached main chain through
complicated motions.

We also recognize that polystyrenes are dynamically heterogeneous (even ignoring ring flippers). Thus, both $T_1\rho(C)$'s and
angular displacement fluctuation parameters must reflect averages
over the entire sample. Of course, the observed $\langle T_1\rho(C)\rangle$ is the
weighted average of all the $T_1\rho(C)$'s present [1]. Since the
$[n_2/n_1]_0^*$ ratio has an approximately linear dependence on total
angular displacement (and is not critically model dependent), the
observed rms fluctuations parameter, θ is also a simple
weighted average. In part, the correlation of Table III
(between $\langle T_1\rho(C)\rangle$ and $\sin^2\theta$) succeeds because we ignore details
of the distributions of motions. Thus, whether polystyrene has
a fraction of mobile rings with the remainder less mobile, or
whether all rings have an intermediate mobility, becomes
immaterial. Both situations result in comparable average values
for $\langle T_1\rho(C)\rangle$ and θ. The correlation of Table III fails to the
extent there remain large-amplitude high-frequency motions which
reduce $[n_2/n_1]_0^*$ but do not contribute to $T_1\rho(C)$, or 5-10 degree
small-amplitude low-frequency motions which can still make
significant contributions to $\langle T_1\rho(C)\rangle$ but have only a minor effect
on $[n_2/n_1]_0^*$.

Literature Cited

1. J. Schaefer, E. O. Stejskal, T. R. Steger, M. D. Sefcik and
   R. A. McKay, Macromolecules, 13, 1121 (1980).
2. J. Schaefer, M. D. Sefcik, E. O. Stejskal and R. A. McKay,
   Macromolecules, 14, 280 (1981).
3. U. Haeberlen, "High Resolution NMR in Solids: Selective
   Averaging," (Adv. Magn. Reson., Suppl. 1), Academic Press,
   New York, 1976.
4. R. K. Hester, J. L. Ackerman, B. L. Neff, and J. S. Waugh,
   Phys. Rev. Letters, 36, 1081 (1976).
5. M. E. Stoll, A. J. Vega, and R. W. Vaughan, J. Chem. Phys.
   65, 1093 (1976).
6. M. G. Munowitz, R. G. Griffin, G. Bodenhausen, and T. H.
   Haung, J. Am. Chem. Soc., 103, 2529 (1981).
7. M. G. Munowitz and R. G. Griffin, J. Chem. Phys., 76, 2848
   (1982).
8. J. Schaefer, R. A. McKay, E. O. Stejskal, and W. T. Dixon,
   J. Mag. Reson., 52, 123 (1983).
9. M. Maricq and J. S. Waugh, J. Chem. Phys., 70, 3300 (1979).
10. J. Herzfeld and A. E. Berger, J. Chem. Phys., 73, 6021
    (1980).
11. J. Schaefer, M. D. Sefcik, E. O. Stejskal, R. A. McKay,
    W. T. Dixon, and R. E. Cais, Macromolecules, March, 1984.
12. A. Abragam, "The Principles of Nuclear Magnetism," Oxford
    University Press, London, 1961, p. 565.

RECEIVED November 18, 1983

Solid State ^{2}H NMR Studies of Molecular Motion

Poly(butylene terephthalate) and Poly(butylene terephthalate)-Containing Segmented Copolymers

L. W. JELINSKI and J. J. DUMAIS

Bell Laboratories, Murray Hill, NJ 07974

A. K. ENGEL

E. I. du Pont de Nemours and Company, Wilmington, DE 19898

Solid state deuterium NMR spectroscopy is used to provide information concerning the motional heterogeneity and homogeneity of segmented co-polyesters containing poly(butylene terephthalate) as the hard segment. The results presented here provide the first clear evidence that there are two distinct motional environments for the hard segments in the co-polyester. One of the environments is identical to that observed in the poly(butylene terephthalate) homopolymer, in which Helfand-type motions about three bonds (Helfand, E. *J. Chem. Phys.* **1971**, *54*, 4651) occur with a correlation time of 7×10^{-6} s at 20°C. The other motional environment is more-nearly isotropic. The residues in the mobile environment are attributed to short blocks of hard segments residing in the soft segment matrix, or to hard segments forming the loop regions in the poly(butylene terephthalate) lamellae. Approximately 10% of the hard segments reside in the mobile environment in the segmented copolymer with 0.87 mole fraction of poly(butylene terephthalate) hard segments.

Poly(butylene terephthalate) has been used as a model for observing motions about three bonds as an isolated motional mode in polymers. Early carbon NMR studies involving carbon NMR relaxation data (1,2) and carbon chemical shift anisotropy considerations (3) showed that the terephthalate residues can be considered static in comparison to the motions exhibited by the alkyl residues. Furthermore, the alkyl portion of poly(butylene terephthalate) contains the shortest sequence that is able to undergo motions about three bonds (4), with the terephthalate groups acting as "molecular anchors" to prevent the longer range motional modes. Poly(butylene terephthalate) is thus ideally set up to undergo three-bond types of motions.

Solid state deuterium NMR spectroscopy was used to study these motions in detail (5), using the selectively labeled polymer:

0097-6156/84/0247-0055$06.00/0
© 1984 American Chemical Society

$$\left[\!\!\begin{array}{c}O\\\parallel\\C\end{array}\!\!-\!\!\bigcirc\!\!-\!\!\begin{array}{c}O\\\parallel\\C\end{array}\!\!-O-CH_2CD_2CD_2CH_2-O\right]_X$$

The temperature-dependent spectra were interpreted in terms of a two-site hop model, in which the deuterons undergo jumps through a dihedral angle of 103°. This type of motion is consistent with gauche-trans conformational transitions. At -88°C these motions appear static on the time scale of the deuterium NMR experiment, and at +85°C the motions are in the fast exchange limit. The rate constants for these motions were obtained from the calculated spectra. An Arrhenius plot of these data show that the apparent activation energy is 5.8 kcal/mol. (Dynamic mechanical data (20 Hz) fall on the Arrhenius plot.) The transitions have an intermediate rate on the deuterium NMR time scale at 20°C, with the correlation time for the motion being 7×10^{-6} s at this temperature.

The solid state deuterium NMR results for the selectively labeled poly(butylene terephthalate) homopolymer have been interpreted in terms of various models for polymer motion (5). The data are consistent with the models proposed by Helfand (6), in which counter rotation occurs about second neighbor parallel bonds. Termed *gauche migration* and *pair gauche production*, these motions are illustrated in Figure 1a and b, respectively. Gauche migration consists of a transition from $g^{\pm}tt$ to $ttg^{\pm}$, and is part of the pathway for pair gauche production. In this latter motional mode, a *ttt* sequence goes to $g^{\pm}tg^{\mp}$. These motional models allow gauche-trans conformational transitions to occur without concomitant large-scale reorientation of the ends of the polymer chain. In addition, they require slightly more than one C-C bond rotational barrier height, and are thus consistent with the apparent activation energy of 5.8 kcal/mol.

Having established the rates and types of motion that occur in the poly(butylene terephthalate) homopolymer (5), it is of interest to extend these studies to the investigation of the types of motions that occur in segmented copolymers that contain poly(butylene terephthalate) as the hard segment. The structure of such a polymer is shown below, where m and n refer to the "hard" and "soft" segments, respectively. (This structure also illustrates the sites of the deuterium labels.)

$$\left[\!\!\begin{array}{c}O\\\parallel\\C\end{array}\!\!-\!\!\bigcirc\!\!-\!\!\begin{array}{c}O\\\parallel\\C\end{array}\!\!-O-CH_2CD_2CD_2CH_2-O\right]_m \cdots \left[\!\!\begin{array}{c}O\\\parallel\\C\end{array}\!\!-\!\!\bigcirc\!\!-\!\!\begin{array}{c}O\\\parallel\\C\end{array}\!\!-O\left\{(CH_2)_4\,O\right\}_{12}\right]_n$$

The poly(butylene terephthalate) "hard" segments and the poly(tetramethyleneoxy) terephthalate "soft" segments are not completely miscible and thus lead to phase separation. However, in contrast to the discrete domain structures found in polystyrene — polybutadiene or polystyrene — polyisoprene, for example, the hard and soft segments of this copolyester are more intimately dispersed. This leads to a hard segment morphology that has been described as "continuous and interpenetrating lamellae." Such a morphology is shown in schematic form in Figure 2. This copolymer thus presents a unique opportunity to study a system in which there is a large interfacial area.

The specific questions to be addressed by solid state deuterium NMR studies of this polymer are the following: (1) Are the motions of the CD_2 groups in the segmented copolymer identical to the motions observed in the poly(butylene terephthalate) homopolymer? (2) What effect do variations in the hard:soft content ratio have on the motions of the CD_2 groups? And (3), is there evidence for hard segments which reside in the soft segment matrix? If so, what per cent of them are in non-poly(butylene terephthalate)-like environments?

EXPERIMENTAL

The selectively labeled poly(butylene terephthalate)-containing segmented copolymers were prepared according to literature methods (7), using 2,2,3,3-d_4-butylene glycol (Merck) as the starting diol. The polymers used in this study correspond to m:n ratios (see the structure, above, for the meaning of m and n) of 24:1 and 7:1. The mole fractions of hard segments are 0.96 and 0.87, respectively, corresponding to 81 and 57 weight per cent of hard segments. The polymers were characterized by thermal measurements and by solution state deuterium and carbon NMR spectroscopy (8). No end groups were observed by carbon spectroscopy, and the deuterium NMR spectra in solution attested to the integrity of the labeling pattern.

The samples for NMR spectroscopy were melted into glass tubes and allowed to cool from the melt. The observed deuterium NMR spectra are reproducible with temperature cycling, thus providing evidence that the thermal history induced by acquiring temperature-dependent spectra of the samples does not greatly affect the properties that we are measuring.

The solid state deuterium NMR spectra were recorded on a home-built spectrometer operating at 55.26 MHz for deuterium (360 MHz for protons). The spectrometer has been described previously (5). Routine spectra were obtained in quadrature using the quadrupolar echo pulse sequence (Figure 3) $(90_{\pm x}-t_1-90_y-t_2)$ (9-11), 4K data points, a digitization rate of 200 ns/point (5 MHz), and a 4.3 μs 90 degree pulse width. Unless otherwise noted, the length of t_1 was generally set at 30 μs. The value of t_2 was set several microseconds shorter than the time needed to start digitization at the top of the quadrupolar echo maximum. After data accumulation, the FID was left-shifted by the correct number of points, so that for each spectrum, the part of the FID which was transformed began at the exact top of the echo. The quadrupolar echo pulse sequence is used to circumvent problems with receiver recovery times. The solid

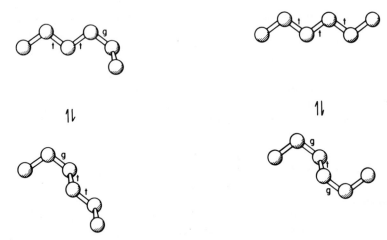

Figure 1. Helfand-type motions about three bonds (6). The segments represented here correspond to the $-OCH_2CD_2CD_2CH_2O-$ portion of the poly(butylene terephthalate) hard segments. These motions are consistent with the solid state deuterium NMR data for poly(butylene terephthalate) (5).

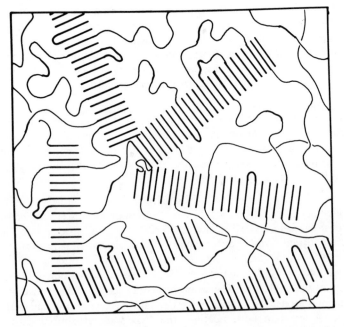

Figure 2. Schematic representation of continuous and interpenetrating hard segment lamellae. The hard segment blocks that are too short to crystallize are shown in the soft segment matrix.

state **deuterium** NMR spectra are very broad (ca. 250 kHz), and thus the free induction decay signal dies away very rapidly. The quadrupolar echo pulse sequence causes the magnetization to refocus while the radio frequency circuits have time to recover from the transmitter pulses.

Inversion-recovery deuterium NMR spectra were obtained by performing a $180°$-τ-$90°$ pulse sequence, followed by the quadrupolar echo sequence (**12**). Spin lattice relaxation times were estimated from the null points in the inversion-recovery spectra.

RESULTS AND DISCUSSION

In Figure 4, the solid state deuterium NMR spectrum for poly(butylene terephthalate) at $20°C$ is compared to the spectra for the segmented copolymers, also obtained at $20°C$. The spectrum of poly(butylene terephthalate) (Figure 4a) can be simulated by assuming the C-D bond jumps between two sites separated by a dihedral angle of $103°$, with a rate constant of 1.4×10^5 s^{-1}. A quadrupolar splitting ($\Delta\nu_q$) of 124 kHz is used for this calculation (**5**). (The calculated deuterium spectrum is obtained by also taking into account pulse power fall-off as a function of frequency (**13**) and the distortions that arise when motions occur during the quadrupolar echo delay time (**14,15**).) Figure 4b shows the deuterium NMR spectrum of the segmented copolymer with 0.96 mole fraction hard segments, and below it in Figure 4c is the spectrum of the segmented copolymer with 0.87 mole fraction hard segments. The spectra of the segmented copolymers are clearly different from the spectrum of poly(butylene terephthalate), and suggest, particularly in the case of the softest segmented copolymer (Figure 4c), that there are at least two motional environments for the hard segments in the copolymers.

Inversion-recovery solid state deuterium NMR spectra can be used to show that the spectrum of the copolymer with 0.87 mole fraction of hard segments is composed of at least two components. Figure 5 shows such spectra. At an inversion-recovery delay time of 50 ms, the broad poly(butylene terephthalate)-like part of the line is almost nulled, yet the sharp component is positive (Figure 5b). At shorter inversion-recovery delay times, the sharp component goes through its null, and the broad component is negative. Inversion-recovery spectra such as these indicate that the solid state deuterium NMR spectra for the segmented copolymers shown in Figures 4b and 4c are composed of two components with different T_1 values. The T_1 of the sharp component in the copolymer with 0.87 mole fraction hard segments is estimated to be 10 ms, and that of the broad component is approximately 60 ms. The difference in these deuterium T_1 values is a factor of six, and indicates that the deuterons that give rise to the broad component are in markedly different motional environments from those that give the sharp line.

It is noteworthy that the sharp component gives a signal with the quadrupolar echo pulse sequence. The observation of a quadrupolar echo can be interpreted to indicate that the constraints on the motion of the deuterons persist for a long time (**16**). Although the deuterium line for the $T_1 = 10$ ms component is sharp, the observation of a quadrupolar echo proves that the line is not isotropically mobile on the deuterium NMR timescale. The sharp component also gives a spectrum with a

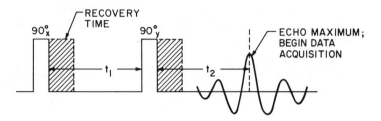

Figure 3. The quadrupole echo pulse sequence. The time t_1 is usually set at 30 μs, and t_2 is approximately 25 μs.

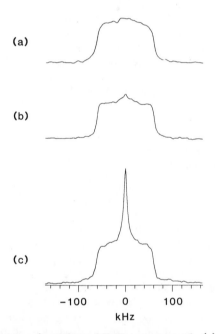

Figure 4. Solid state deuterium NMR spectra of (a) poly(butylene terephthalate); (b) segmented co-polyester containing 0.96 mole fraction of poly(butylene terephthalate) hard segments; and (c) segmented copolymer containing 0.87 mole fraction hard segments. (See text for specific deuterium labeling patterns.) All spectra were obtained with the quadrupole echo pulse sequence at 20°C and 55.26 MHz, using 30 μs at the t_1 quadrupole echo delay time.

standard 90° − τ pulse sequence using a 30 μs receiver dead time (spectrum not shown). The width of this line at half-height is approximately 5 kHz.

The spectra shown in Figure 5 clearly indicate that there are two distinct motional environments in the segmented copolymers. The broad component of the spectrum arises from the majority of the hard segments which show motional characteristics similar to that of poly(butylene terephthalate). The sharp component is attributed to hard segments which reside in the soft segment matrix, either by virtue of being very short blocks, or because they form loops on the surfaces of the hard segment lamellae.

It is of interest to estimate the amount of hard segment which gives the sharp line, and further, to determine if the poly(butylene terephthalate)-like broad line represents deuterons which undergo motions which are identical to those observed in the poly(butylene terephthalate) homopolymer. These points are addressed for the copolymer containing 0.87 mole fraction hard segments, by the solid state deuterium NMR spectra shown in Figure 6.

In Figure 6a is shown the quadrupole echo deuterium NMR spectrum of the segmented copolymer with 0.87 mole fraction of hard segments. Adjacent to it in Figure 6b, is the corresponding calculated spectrum. The dashed line represents the sum of the sharp and broad components. The sharp line is Lorentzian, and the broad line is Gaussian. The broad line is calculated assuming a quadrupolar splitting $(\Delta\nu_q)$ of 124 kHz, a two-site dihedral hop angle of 103°, and a rate constant for the hop of 1.4×10^5 s^{-1}. Corrections have been made for pulse power fall-off (13) and for distortions that arise when motions occur during the quadrupolar echo pulse sequence (14,15). This calculation indicates that the sharp and broad components are present in an approximately 1:10 ratio.

The spectrum shown in Figure 6c illustrates that the motions of the CD$_2$ groups of poly(butylene terephthalate) in the segmented copolymer are identical to the motions in the homopolymer. This spectrum is a difference spectrum, obtained by subtracting the spectrum of the poly(butylene terephthalate) homopolymer (Figure 4a) from the spectrum of the segmented copolymer (Figure 4c). The difference spectrum (Figure 6c) shows an excellent null for the broad part of the deuterium NMR pattern. This result indicates that the motions of approximately 90% of the hard segment CD$_2$ deuterons in the segmented copolymer are well-represented by the motions observed for the poly(butylene terephthalate) homopolymer.

It is emphasized that the semi-quantitative treatment above provides only an estimate of the amounts of hard segments in the two motional environments. A more-quantitative answer is probably unlikely, due to the quadrupolar echo pulse sequence necessary to obtain the data. The quadrupolar echo pulse sequence preserves the inhomogeneously broadened part of the solid state deuterium NMR pattern. Homogeneous T$_2$ effects will cause some signal loss during the quadrupolar echo delay times. In comparing sharp and broad components such as those shown in Figure 6b, it is also necessary to know that both components have approximately the same homogeneous T$_2$.

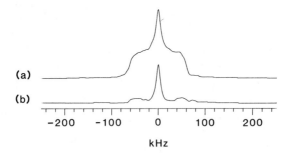

Figure 5. Inversion-recovery solid state deuterium NMR spectra of the segmented copolymer containing 0.87 mole fraction hard segments. The spectra were obtained with a $180°-t_3-90°_{\pm x}-t_1-90°_y-t_2$ pulse sequence. The spectrum in (b) was obtained with a t_3 of 50 ms.

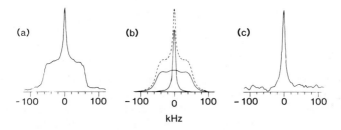

Figure 6. Experimental (a), calculated (b), and difference (c) solid state deuterium NMR spectra for the segmented copolymer with 0.87 mole fraction of hard segments. The spectrum in (a) was obtained at 55.26 MHz and 20°C, using the quadrupole echo pulse sequence. The dashed line in (b) represents the sum calculated for the broad and narrow components in a respective 10:1 ratio. (See text for details of the calculation.) The spectrum in (c) is the difference spectrum obtained by subtracting the spectrum of poly(butylene terephthalate) (Figure 4a) from the segmented copolymer spectrum (Figure 6a).

Homogeneous T_2 effects can be studied by obtaining quadrupolar echo NMR spectra as a function of the quadrupolar echo delay times, t_1 and t_2. Such experiments have been performed for all samples. The results indicate that the observed lineshapes are not markedly dependent upon the quadrupolar echo delay times, although signal intensity clearly is lost as the delay times are increased. Figure 7 shows a plot of the relative signal intensity versus quadrupolar echo delay time for the solid state deuterium NMR spectra of the segmented copolymer containing 0.87 mole fraction hard segments. This figure shows that the relative intensity is a linear function of the quadrupolar echo delay times. When extrapolated back to zero time, Figure 7 illustrates that approximately 25% of the signal intensity is lost at the usual quadrupolar echo delay time of 30 μs. However, the lack of distortion in these spectra indicates that representative patterns are obtained at a quadrupolar echo delay time of 30 μs.

SUMMARY

The results presented here show that there are two distinct motional environments for the "central" deuterons of poly(butylene terephthalate) in the segmented co-polyesters containing poly(butylene terephthalate) as the hard segment. One of the motional environments is identical to that observed in the poly(butylene terephthalate) homopolymer, in which Helfand-type motions (6) about three bonds occur with a correlation time of 7×10^{-6} s at 20°C. The T_1 for these deuterons is *ca.* 60 ms at 20°C. Approximately 90% of the hard segments reside in these organized lamellar environments in the segmented copolymer with 0.87 mole fraction hard segments.

The other 10% of the hard segments in this copolymer undergo motions that are more-nearly isotropic. The T_1 for the "central" deuterons in the mobile regions is approximately 10 ms at 20°C. The mobile residues are attributed to short blocks of hard segments residing in the soft segment matrix, or to hard segments forming the loop regions of the poly(butylene terephthalate) lamellae. As the mole fraction of hard segments is increased from 0.87 to 0.96, a lower per cent of the hard segments are found to reside in mobile environments.

Taken together, these solid state deuterium NMR experiments provide otherwise unobtainable answers to questions concerning motional homogeneity and heterogeneity in these segmented copolymer systems.

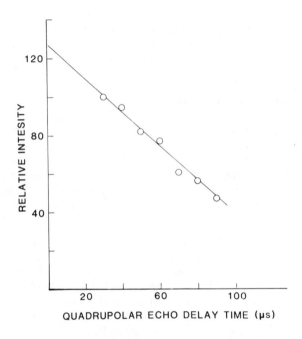

Figure 7. A plot of the relative intensity of the quadrupole echo deuterium NMR signal versus the echo delay time, t_1, for the segmented copolymer containing 0.87 mole fraction hard segments. The line is extrapolated back to zero time.

ACKNOWLEDGEMENT

We are grateful to Dr. F. A. Bovey for his interest and encouragement in this project.

LITERATURE CITED

1. Jelinski, L. W.; Dumais, J. J. *Polym. Prepr., Am. Chem. Soc. Div. Polym. Chem.* **1981**, *22* (2), 273.

2. Jelinski, L. W.; Dumais, J. J.; Watnick, P. I.; Engel, A. K.; Sefcik. M. D. *Macromolecules* **1983**, *16*, 409.

3. Jelinski, L. W. *Macromolecules* **1981**, *14*, 1341.

4. Schatzki, T. F. *Polym. Prepr., Am. Chem. Soc. Div. Polym. Chem.* **1965**, *6*, 646.

5. Jelinski, L. W.; Dumais, J. J.; Engel, A. K. *Macromolecules* **1983**, *16*, 492.

6. Helfand, E. *J. Chem. Phys.* **1971**, *54*, 4651.

7. Wolfe, J. R., Jr. *Polym. Prepr., Am. Chem. Soc. Div. Polym. Chem.* **1978**, *19* (1), 5.

8. Jelinski, L. W.; Schilling F. C.; Bovey, F. A. *Macromolecules* **1981**, *14*, 581.

9. Davis, J. H.; Jeffrey, K. R.; Bloom, M.; Valic, M. I.; Higgs, T. P. *Chem. Phys. Lett.* **1976**, *42*, 390.

10. Blinc, R.; Rutar, V.; Seliger, J.; Slak, J.; Smolej, V. *Chem. Phys. Lett.* **1977**, *48*, 576.

11. Hentschel, R.; Spiess, H. W. *J. Magn. Resonance* **1979**, *35*, 157.

12. Batchelder, L. S.; Niu, C. H., Torchia, D. A. *J. Am. Chem. Soc.* **1983**, 0000.

13. Bloom, M.; Davis, J. H.; Valic, M. I. *Can. J. Phys.* **1980**, *58*, 1510.

14. Spiess, H. W.; Sillescu, H. *J. Magn. Resonance* **1981**, *42*, 381.

15. Mehring, M. In "NMR-Basic Principles and Progress", Diehl, P.; Fluck, E.; Kosfeld, R.; Eds.; Springer-Verlag: New York, *11*, **1976**.

16. Hentschel, D.; Sillescu, H.; Spiess, H. W. *Macromolecules* **1981**, *14*, 1605.

RECEIVED December 10, 1983

Spin Relaxation and Local Motion in a Dissolved Aromatic Polyformal

M. F. TARPEY, Y.-Y. LIN, and ALAN ANTHONY JONES

Jeppson Laboratory, Department of Chemistry, Clark University, Worcester, MA 01610

P. T. INGLEFIELD

Department of Chemistry, College of the Holy Cross, Worcester, MA 01610

Carbon-13 and proton spin-lattice relaxation times are reported for 10 wt% solutions of a dissolved aromatic polyformal. The relaxation times for both nuclei were determined at two Larmor frequencies and as a function of temperature from 0 to 120°C. These relaxation times are interpreted in terms of segmental motion and anisotropic internal rotation. Segmental correlation functions by both Jones and Stockmayer, and Weber and Helfand were used to interpret the data. Internal rotation is described by the usual Woessner approach, and restricted rotational diffusion, by the Gronski approach. Both segmental correlation functions lead to similar results; but, relative to the analogous polycarbonate, single bond conformational transitions are more frequent in the polyformal. The phenyl groups in the backbone undergo segmental rearrangements and internal anisotropic rotation at comparable rates. Motion in the formal linkage is described by the same segmental correlation times plus restricted rotational diffusion about an axis between the oxygens of the formal group. The interpretation at the formal link based on restricted rotational diffusion is discussed in terms of the conformations likely in the link which are commonly referred to as the anomeric effect. The choice of the axis of restricted rotation in the formal unit is only an approximation of the result of anisotropic single bond conformational transitions occurring within that unit.

0097–6156/84/0247–0067$06.00/0

Spin relaxation in dilute solution has been employed to characterize local chain motion in several polymers with aromatic backbone units. The two general types examined so far are polyphenylene oxides (1-2) and aromatic polycarbonates (3-5); and these
two types are the most common high impact resistant engineering
plastics. The polymer considered in this report is an aromatic
polyformal (see Figure 1) where the aromatic unit is identical to
that of one of the polycarbonates. This polymer has a similar dynamic mechanical spectrum to the impact resistant polycarbonates (6) and is therefore an interesting system for comparison of
chain dynamics.

In addition, the formal unit itself offers a new opportunity
for monitoring chain motion relative to the polycarbonates since
the carbonate unit contains no protons. The spin-lattice relaxation times, T_1's of all protons and all carbons with directly
bonded protons are reported for the polyformal. Also the carbon
and proton T_1's are measured at two different Larmor frequencies
to expand the frequency range covered by the study.

In addition to determining the time scales for several local
motions in polyformal, two different interpretational models for
segmental motion will be employed. An older model by Jones and
Stockmayer (7), based on the action of a three bond jump on a
tetrahedral lattice is compared with a new model by Weber and Helfand (8), based on computer simulations of polyethylene type
chains. These two models for segmental motion have been compared
before (5) for two polycarbonates but somewhat different results
are seen in the polyformal interpretation.

Experimental

High molecular weight samples of the polyformal were kindly supplied by General Electric. The structure of the repeat unit is
shown in Figure 1 as well as the structure of a partially deuterated form which was synthesized (9) to reduce proton cross-
relaxation. A 10 weight percent solution of the polymer in deuterated tetrachloroethane was prepared in an NMR tube, subjected
to five freeze, pump, thaw cycles and sealed.

Spin lattice relaxation measurements were conducted on two
spectrometers with a standard $\pi-\tau-\pi/2$ pulse sequence. The 30 and
90 MHz proton measurements as well as the 22.6 MHz carbon-13 measurements were made on a Bruker SXP 20-100. The 250 MHz proton and
62.9 carbon-13 measurements were made on a Bruker WM-250.

Results

Spin lattice relaxation times are calculated from the return of
the magnetization to equilibrium using a linear and non-linear
least squares analysis of the data. The two analyses yield T_1
values within 10% of each other and average values are reported.
No evidence of cross-relaxation or cross-correlation were observed

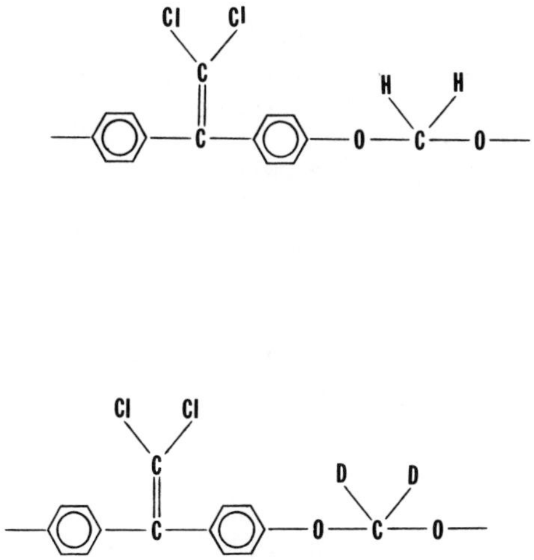

Figure 1. Structure of the polyformal repeat unit and the partially deuterated analogue.

in the form of nonlinear behavior when the return of the magnetization was plotted in the standard from $\ln(A_\infty - A_\tau)$ versus τ. The
presence of cross-relaxation in the proton data was further check-
ed by comparing the phenyl proton T_1 in the partially deuterated
analogue with the phenyl T_1 in the fully protonated polymer. The
phenyl proton T_1 is about 10% longer in the deuterated polymer indicating a small amount of cross-relaxation though the 10% change
is essentially the same as the experimental uncertainty. A 10%
uncertainty is placed on all T_1 values, reflecting error contributions from concentration, temperature and pulse widths as well as
random fluctuations in intensity. Table I contains proton and
carbon-13 T_1's as a function of temperature and Larmor frequency.

Interpretation

The standard relationships between T_1's and spectral densities J's
are employed. For carbon-13, the expressions are (4)

$$1/T_1 = W_0 + 2W_{1C} + W_2$$

$$W_0 = \Sigma_j \gamma_C^2 \gamma_H^2 h^2 J_1(\omega_0)/20r_j^6$$

$$W_{1C} = \Sigma_j 3\gamma_C^2 \gamma_H^2 h^2 J_1(\omega_C)/40r_j^6 \qquad (1a)$$

$$W_2 = \Sigma_j 3\gamma_C^2 \gamma_H^2 h^2 J_2(\omega_2)/10r_j^6$$

$$\omega_0 = \omega_H - \omega_C \qquad \omega_2 = \omega_H + \omega_C$$

and for protons it is

$$1/T_1 = \Sigma_j (9/8)\gamma^4 h r_j^{-6}[(2/15)J_1(\omega_H) + (8/15)J_2(2\omega_H)] \qquad (1b)$$

The internuclear distances employed are 1.095 Å for the phenyl C-H
distance, 1.125 Å for the formal C-H distance, 2.4 Å for the 2-3
phenyl proton distance, and 1.75 Å for the formal proton - proton
distance. The 2-3 phenyl proton distance used here is comparable
to the distance of 2.41 Å used in the polycarbonate interpretations. The choice of 2.4 Å is based on the phenyl proton T_1 minimum and the slightly smaller value is confirmed by a larger Pake
doublet splitting observed in the solid state spectrum of the
phenyl protons in the partially deuterated analogue (10).
      Expressions for the spectral density can be developed from
models for local motion in randomly coiled chains. Two general
types of local motion will be considered, and they are segmental
motion and anisotropic rotation. Segmental motion itself will be

Table I: Spin-Lattice Relaxation times (ms)

°C	Phenyl Protons		Protonated Phenyl Carbons		Formal Protons		Formal Carbons	
	90 MHz	30 MHz	62.9 MHz	22.6 MHz	250 MHz	90 MHz	62.9 MHz	22.6 MHz
0	548	153	137	72	335	129	81	36
20	499	189	174	106	282	114	91	52
40	522	274	243	156	268	133	115	82
60	628	432	377	349	295	198	158	123
80	794	553	543	448	350	258	229	193
100	1168	763	798	679	467	320	339	275
120	1411	939	1113	936	628	433	403	377

described in two ways. The first description is derived from the
action of a three bond jump on a tetrahedral lattice ($\underline{7}$) and the
second is developed from consideration of computer simulations of
backbone transitions in polyethylene chains ($\underline{8}$). Anisotropic rotation can also be characterized in several ways. It can be described as jumps between two minima ($\underline{11}$), jumps between three minima ($\underline{12}$) or stochastic diffusion ($\underline{12}$).

In the three bond jump model for segmental motion there are
two parameters. The time scale is set by the harmonic average
correlation time, τ_h and the effective distribution of correlation
times is set by the number of coupled bonds m. The sharp cut off
of coupling solution of the three bond jump model is employed
here. The composite spectral density for internal rotation by
jumps or stochastic diffusion plus segmental motion by three bond
jump is

$$J_i(\omega_i) = 2\sum_{k=1}^{s} G_k \frac{A\tau_{k0}}{1 + \omega_i^2\tau_{k0}^2} + \frac{B\tau_{bk0}}{1 + \omega_i^2\tau_{bko}^2} + \frac{C\tau_{ck0}}{1 + \omega_i^2\tau_{ck0}^2}$$

$$\tau_{k0}^{-1} = \tau_k^{-1}$$

$$\tau_k = W\lambda_k \qquad s = (m + 1)/2$$

$$\lambda_k = 4 \sin^2[(2k - 1)\pi/2(m + 1)]$$

$$\tau_h^{-1} = 2W$$

$$G_k = 1/s + (2/s) \sum_{q=1}^{s-1} \exp(-\gamma q) \cos [(2k - 1)\pi q/2s] \qquad (2)$$

$$\gamma = \ln 9$$

$$A = (3 \cos^2 \Delta - 1)^2/4$$

$$B = 3(\sin^2 2\Delta)/4$$

$$C = 3(\sin^4 \Delta)/4$$

for stochastic diffusion

$$\tau_{bk0}^{-1} = \tau_k^{-1} + \tau_{ir}^{-1}$$

$$\tau_{ck0} = \tau_k^{-1} + (\tau_{ir}/4)^{-1}$$

for a threefold jump

$$\tau_{bk0}^{-1} = \tau_{ck0}^{-1} = \tau_k^{-1} + \tau_{ir}^{-1}$$

The angle Δ is between the internuclear vector and the axis of internal rotation.

The three bond jump segmental motion description can also be combined with a description of restricted anisotropic rotational diffusion (13-14). In this case, the composite spectral density equation is

$$J_i(\omega_i) = 2\sum_{k=1}^{s} G_k \frac{A\tau_{k0}}{1 + \omega_i^2 \tau_{k0}^2} +$$

$$\frac{B}{\ell^2} \left\{ [(1 - \cos \ell)^2 + \sin^2 \ell] \frac{\tau_{k0}}{1 + \omega_i^2 \tau_{k0}^2} + \right.$$

$$\frac{1}{2} \sum_{n=1}^{\infty} \left[\left\{ \frac{[1 - \cos(\ell - n\pi)]}{(1 - \frac{n\pi}{\ell})} + \frac{[1 - \cos(\ell + n\pi)]}{(1 + \frac{n\pi}{\ell})} \right\}^2 + \right.$$

$$\left. \left\{ \frac{\sin(\ell - n\pi)}{(1 - \frac{n\pi}{\ell})} + \frac{\sin(\ell + n\pi)}{(1 + \frac{n\pi}{\ell})} \right\}^2 \right] \frac{\tau_{nk0}}{1 + \omega_i^2 \tau_{nk0}^2} \left. \right\} + \qquad (3)$$

$$\frac{C}{2\ell^2} \left\{ \frac{1}{2} [(1 - \cos 2\ell)^2 + \sin^2 2\ell] \frac{\tau_{k0}}{1 + \omega_i^2 \tau_{k0}^2} + \right.$$

$$\sum_{n=1}^{\infty} \left[\left\{ \frac{[1 - \cos(2\ell - n\pi)]}{(2 - \frac{n\pi}{\ell})} + \frac{[1 - \cos(2\ell + n\pi)]}{(2 + \frac{n\pi}{\ell})} \right\}^2 + \right.$$

$$\left. \left\{ \frac{\sin(2\ell - n\pi)}{(2 - \frac{n\pi}{\ell})} + \frac{\sin(2\ell + n\pi)}{(2 + \frac{n\pi}{\ell})} \right\}^2 \right] \frac{\tau_{nk0}}{1 + \omega_i^2 \tau_{nk0}^2} \left. \right\} +$$

where $\dfrac{1}{\tau_{k0}} = \dfrac{1}{\tau_k}$

$$\frac{1}{\tau_{nk0}} = \frac{1}{\tau_k} + \lambda_n \qquad \text{and}$$

$$\lambda_n = \left(\frac{n\pi}{\ell}\right)^2 D_{ir}$$

The new parameters for restricted anisotropic rotational diffusion
are the angular amplitude over which rotation diffusion occurs, ℓ,
and the rotational diffusion constant for restricted anisotropic
rotational diffusion, D_{ir}.

A second description of segmental motion can be combined with
the various types of internal anisotropic internal rotation.
Weber and Helfand (8) characterize segmental motion in terms of a
correlation time for single conformational transitions, τ_0, and a
correlation time for cooperative conformational transitions, τ_1.
This model has been applied to nuclear spin relaxation before (5)
and the form of the spectral density for a composite segmental
motion and anisotropic internal rotation is written

$$J_i(\omega_i) = AJ_{ia}(\tau_0, \tau_1, \omega_i) + BJ_{ib}(\tau_{b0}, \tau_1, \omega_i) + CJ_{ic}(\tau_{c0}, \tau_1, \omega_i)$$

$$A = (3 \cos^2 \Delta - 1)^2/4$$

$$B = 3 (\sin^2 2\Delta)/4 \tag{4}$$

$$C = 3(\sin^4 \Delta)/4$$

for stochastic diffusion

$$\tau_{b0}^{-1} = \tau_0^{-1} + \tau_{ir}^{-1}$$

$$\tau_{c0}^{-1} = \tau_0^{-1} + (\tau_{ir}/4)^{-1}$$

for a three bond jump

$$\tau_{b0}^{-1} = \tau_{c0}^{-1} = \tau_0^{-1} + \tau_{ir}^{-1}$$

The form of J_{ia}, J_{ib} and J_{ic} is the same as J_{ij} given below with
τ_0 replaced by τ_0, τ_{b0} and τ_{c0} respectively.

$$J_{ij}(\omega_i) = 2\{[(\tau_0^{-1})(\tau_0^{-1} + 2\tau_1^{-1}) - \omega_i^2]^2 + [2(\tau_0^{-1} + \tau_1^{-1}\omega_i)]^2\}^{-1/4}$$

$$x \cos [1/2 \arctan(2(\tau_0^{-1} + \tau_1^{-1})\omega_i/\tau_0^{-1}(\tau_0^{-1} + 2\tau_1^{-1}) - \omega_i^2]$$

This description of segmental motion can also be combined with
restricted anisotropic rotational diffusion

$$J_i(\omega_i) = AJ_i^{01}(\omega_i) +$$

$$\frac{B}{\ell^2}\left\{[(1 - \cos \ell)^2 + \sin^2 \ell]\, J_i^{01}(\omega_i) + \right.$$

$$\frac{1}{2}\sum_{n=1}^{\infty}\left[\left\{\frac{[1 - \cos(\ell - n\pi)]}{(1 - \frac{n\pi}{\ell})} + \frac{[1 - \cos(\ell + n\pi)]}{(1 + \frac{n\pi}{\ell})}\right\}^2 + \right.$$

$$\left.\left\{\frac{\sin(\ell - n\pi)}{(1 - \frac{n\pi}{\ell})} + \frac{\sin(\ell + n\pi)}{(1 + \frac{n\pi}{\ell})}\right\}^2\right]\, J_i^{\lambda_n}(\omega_i)\right\} +$$

$$\frac{C}{2\ell^2}\left\{\frac{1}{2}[(1 - \cos 2\ell)^2 + \sin^2 2\ell]\, J_i^{01}(\omega_i) + \right.$$

$$\sum_{n=1}^{\infty}\left[\left\{\frac{[1 - \cos(2\ell - n\pi)]}{(2 - \frac{n\pi}{\ell})} + \frac{[1 - \cos(2\ell + n\pi)]}{(2 + \frac{n\pi}{\ell})}\right\}^2 + \right.$$

$$\left.\left\{\frac{\sin(2\ell - n\pi)}{(2 - \frac{n\pi}{\ell})} + \frac{\sin(2\ell + n\pi)}{(2 + \frac{n\pi}{\ell})}\right\}^2\right]\, J_i^{\lambda_n}(\omega_i)\right\}\Bigg] \qquad (5)$$

where

$$J_i^{01}(\omega_i) = \{[\tau_0^{-1}(\tau_{01}^{-1} + \tau_1^{-1}) - \omega_i^2]^2 +$$

$$(2\,\tau_{01}^{-1}\,\omega_i)^2\}^{-1/4} \times \cos\left[\frac{1}{2}\arctan\frac{2\tau_{01}^{-1}\,\omega_i}{\tau_0^{-1}(\tau_{01}^{-1} + \tau_1^{-1}) - \omega_i^2}\right]$$

$$J_i^{\lambda_n}(\omega_i) = \{[(\tau_0^{-1} + \lambda_n)(\tau_{01}^{-1} + \tau_1^{-1} + \lambda_n) - \omega_i^2]^2\}^{-1/4}$$

$$\times \cos\left[\frac{1}{2}\arctan\frac{2(\tau_{01}^{-1} + \lambda_n)\,\omega_i}{(\tau_0^{-1} + \lambda_n)(\tau_{01}^{-1} + \lambda_n) - \omega_i^2}\right]$$

$$\tau_{01}^{-1} = \tau_0^{-1} + \tau_1^{-1}$$

$$\lambda_n = \left(\frac{n\pi}{\ell}\right)^2 D_{ir}$$

All of the terms have been defined in eqs. 2-4.

To apply the models to the interpretation of the data, the
approach developed for the polycarbonates will be followed. The
phenyl proton T_1's are interpreted first in terms of segmental
motion. For these protons, the dipole-dipole interaction is
parallel to the chain backbone and therefore relaxed only by segmental motion. In the three bond jump model the parameters τ_h and
m are adjusted to account for phenyl proton data, and in the
Weber-Helfand model the parameters τ_0 and τ_1 are adjusted. Table
II contains the three bond jump parameters, and Table III, the
Weber-Helfand model parameters. Both models can simulate the data
within 10% which is equivalent to the experimental error.

Phenyl group rotation can be characterized from the phenyl
carbon T_1's by assuming the segmental description developed from
the proton data (5). Either segmental model can be used, and the
corresponding correlation times for internal rotation of the phenyl group by stochastic diffusion, τ_{irp}'s, are displayed in Table
II and Table III. Again both approaches match the observed
carbon-13 data within the 10% uncertainty.

Table II: Phenyl Group Motion Simulation Parameters Using the
Three Bond Jump Model

°C	m	τ_h (ns)	τ_{irp} (ns)
0	1	2.69	1.85
20	1	1.30	1.19
40	1	0.79	0.73
60	1	0.49	0.299
80	3	0.180	0.247
100	5	0.080	0.192
120	7	0.049	0.145
E_a(kJ/mole)		30	20
τ_∞ x 10^{14} (s)		0.59	30
Correlation Coefficient		0.99	0.99

Table III: Phenyl Group Motion Simulation Parameters Using the Weber-Helfand Model

°C	τ_1 (ns)	τ_0 (ns)	τ_{irp} (ns)
0	3.80	6.01	2.15
20	1.89	3.7	1.15
40	1.09	2.34	0.72
60	0.49	2.00	0.280
80	0.259	1.99	0.240
100	0.142	1.86	0.170
120	0.070	1.70	0.150
E_a(kJ/mole)	30	9	21
$\tau_\infty \times 10^{14}$ (s)	1.04	97×10^2	20
Correlation Coefficient	0.99	0.94	0.99

Now the interpretation diverges from the polycarbonate pattern as the formal group is considered. As mentioned, the structural analogue to the formal group in the polycarbonate is the carbonate group, and the latter cannot be directly studied by solution spin relaxation studies since it has no directly bonded protons. If the formal is first viewed independently from the phenyl group data, one might attempt to employ segmental motion descriptions alone since the formal group lies in the backbone. Pursuing this approach, both the three bond jump and the Weber-Helfand models were applied to simulate the proton and carbon-13 data in Table I. Neither model is able to account for the data, with systematic discrepancies up to 70% in both attempts. The largest discrepancies occur at low temperatures with only somewhat better simulations possible at higher temperatures.

In one sense it is reassuring to determine that models for segmental motion cannot account for all data sets. On the other hand, it is still desirable to develop some description of motion which will account for the data at hand, since the failure to simulate implies some potentially interesting informational content. The successful phenyl group interpretation can assist the effort to account for the formal data. The segmental motion descriptions applied to the phenyl proton data are based on isotropic averaging of the dipole-dipole interactions by the segmental motion. One could assume that the same segmental motion description occurring at the phenyl groups also occurs at the formal group since both groups are adjacent in the backbone. If this assumption is made, some additional motion must be considered to match the observed formal relaxation times. In the context of the models being applied, the added motion could be a anisotropic rotation or restricted rotation. For the formal group, the first guess is rota-

tion or restricted rotation about the C–O axis. This would be a
single backbone conformational transition occurring as an anisotropic motion on top of the segmental motion of say the Weber-
Helfand model determined from the phenyl proton data. Complete
anisotropic rotation about the C–O bond adequately accounts for
the higher temperature data, but fails to simulate the lower temperature data by about 40%. A restricted rotation model at lower
temperatures is also not able to simulate the observed T_1's though
it comes closer. Adding a rotation or restricted rotation about
the C–O axis to the three bond jump model is equally unsuccessful
as might be expected since so far the three bond jump and Weber-
Helfand model have paralled each other.

The next motion considered is rotation or restricted rotation
of the OCH_2O unit about the O–O axis of the unit. The initial
logic here was that the larger aromatic groups were slower moving
anchors and the formal group was anisotropically rotating relative
to the two oxygens which were the connections to the more sluggish
phenyl groups. At higher temperatures, complete anisotropic rotation about the O–O axis in addition to a segmental motion description using the Weber-Helfand model developed from the phenyl proton data accounted for the formal data but discrepancies of 30%
still remained at lower temperatures. The lower temperature data
could be accounted for by allowing for incomplete anisotropic rotational diffusion about the O–O axis in addition to segmental motion. With complete rotation at higher temperatures and restricted rotation at lower temperatures, all formal proton and carbon-13
data can be simulated within the experimental uncertainty of the
T_1's. The anisotropic rotation simulation parameters are reported
in Table IV for the case where segmental motion is characterized
with the Weber-Helfand model on the basis of the phenyl proton
data. A substitution of the three bond jump model for the Weber-
Helfand model leads to nearly the same results.

Table IV: Formal Group Simulation Parameters Using the Weber-
Helfand Model[a]

°C	ℓ	$D_{ir} \times 10^{-10}$ (s^{-1})
0	86	0.100
20	119	0.110
40	164	0.130
60	360	0.100
80	360	0.160
100	360	0.210
120	360	0.230

(a) The values of τ_1 and τ_0 reported in Table III are used here
     as well as the parameters listed.

Discussion

As the first point, the dynamics of the phenyl group in the polyformal can be considered. Motional descriptions from the two segmental models can be compared as they have been before for the
polycarbonates (5). In the three bond jump model the primary
parameter is the harmonic mean correlation time, τ_h; and in the
Weber-Helfand model the primary parameter is the correlation time
for cooperative backbone transitions, τ_1. At the lower temperatures studied, τ_0 plays an increasing role in the Weber-Helfand
model but τ_1 is still the major factor. This is an interesting
point in itself since cooperative transitions were also found to
predominate when the Weber-Helfand model was applied to the polycarbonates. Here in the polyformal, single bond conformational
transitions do play a larger role; and this can be seen in the
three bond jump model as well by the drop of m to 1 at lower temperatures. Since τ_1 and τ_h are both measures of the time scale
for cooperative motions, it is interesting to note that the
Arrhenius summaries of the two correlation times in Tables II and
III are very similar. This similarity, taken together with the
domination of cooperative transitions in the interpretations, supports the utility of both models though the Weber-Helfand model is
developed from a more detailed analysis of chain motion.
    One interesting difference between the Weber-Helfand inter-
pretation of the polyformal and the polycarbonates is the relative
apparent activation energies for τ_1 and τ_0. For the polycarbonates, the activation energies for τ_0 and τ_1 were about the
same (5) as would be expected if the cooperative transitions
occurred sequentially as opposed to simultaneously (15-17). For
the polyformal, the activation energy for the cooperative process
is much higher than for the single transitions which is more in-
dicative of simultaneous cooperative transitions such as a crank-
shaft. Since the single transitions are minor processes in both
the polycarbonates and to a lesser extent in the polyformal,
dwelling on the activation energy differences may be risky.
    It is worth noting that the description of phenyl group rota-
tion is not significantly influenced by changing descriptions of
segmental motion. This too supports the utility of both models
and the validity of the general analysis of local motion for phenyl groups as being divided between segmental motion and internal
rotation.
    Segmental motion and phenyl group rotation in the polyformal
can be compared to that of the polycarbonates. Relative to the
analogous Chloral polycarbonate (5), the cooperative segmental motion in the polyformal is similar in general time scale but has a
significantly higher activation energy. Phenyl group rotation in
the polyformal and the polycarbonate are nearly identical. This
suggests phenyl group rotation is a very localized process not
greatly influenced by replacing the carbonate link with a formal
link. On the other hand, it is hard to imagine phenyl group rota-

tion as a simple process within the bisphenol unit since MNDO calculations (18) indicate a high barrier within this unit.

Another interesting point about phenyl group rotation in the
polyformal and polycarbonates is that it is best modeled in solution as stochastic diffusion rather than two fold jump (π flips).
In solid BPA polycarbonate, both deuterium (19) and carbon-13 (20)
lineshape analysis point to two fold jumps or π flips as the primary process. Calculations by Tonelli (21-22) also point to low
barriers to phenyl group rotation for isolated BPA chains. If the
intramolecular barrier for phenyl group rotation is indeed low as
indicated by the solution studies and the calculations, the change
to a higher barrier (6,18) and π flips in the solid must reflect
intermolecular interactions. This is indeed plausible since the
new conformation following a π flip in the solid requires no
change in the surroundings (no change in free volume) yet the surroundings could provide an appreciable barrier to the transition.

As mentioned, the formal link provides new dynamic information relative to the polycarbonates where no detailed analysis of
the carbonate unit is possible. In the interpretation, a rather
complex description is required to account for the formal relaxation data. According to the interpretation, the formal group undergoes segmental motion as determined at the phenyl group plus
anisotropic rotation about the oxygen-oxygen axis of the formal
group. At low temperatures this anisotropic rotation is described
as restricted rotational diffusion. The main question is whether
there is any physical sense to such a picture. Since the segmental motion is somewhat cooperative and the phenyl group is adjacent, it seems reasonable to assume that this motion extends over
both the phenyl and formal groups. The real question is the anisotropic restricted rotation. To pursue this aspect, conformational energy maps of dimethoxymethane were reviewed (23-24). The
lowest conformations are gg' and g'g and this unusual situation
relative to polyethylene chains is commonly called the anomeric
effect. Each of these conformations has two conformations which
are only 4kJ higher in energy. The tg' and gt conformations are
energetically near the gg' conformation and the tg and g't conformations are energetically near the g'g conformation. The g'g', gg
and tt states are considerably higher in energy. The most facile
conformational changes from the lowest states could be represented
by

$$tg' = gg' = gt$$

$$g't = g'g = tg$$

(6)

At lower temperatures where a given formal unit is likely to be
either gg' or g'g, the transitions represented by eq. 6 would result in restricted rotational averaging. This would generally
agree with the results obtained from the simulation of the formal
relaxation data from 0 to 40 degrees where the angular amplitude

of restricted rotation, ℓ, ranges from 86 to 164 degrees. At higher temperatures populations in states other than gg' or g'g would become larger allowing for the more common occurrence of conformational changes other than those listed in eq. 6. This would result in effectively complete rotation in agreement with the simulation from 60 to 120 degrees.

These arguments would account for the shift from restricted rotation to complete anisotropic rotation, but why is the choice of the oxygen-oxygen axis made? In fact, it can only be a rough approximation, since the ends of the formal group must move during these conformational changes. The time scale for the formal group conformational changes are only somewhat more rapid relative to the time scale of segmental motion and phenyl group rotation, so phenyl groups are only somewhat sluggish with respect to the formal group. A more detailed and accurate model for the formal group motion could be undertaken but the data in hand do not warrant it. The present picture points to single conformational transitions at the formal group which result in only partial spatial averaging of dipolar interactions at lower temperatures.

Acknowledgments

The research was carried out with financial support of National Science Foundation Grant DMR-790677, of National Science Foundation Equipment Grant No. CHE 77-09059, of National Science Foundation Grant No. DMR-8108679, and of the U.S. Army Research Office Grant DAAG 29-82-G-0001.

Literature Cited

(1) A.A. Jones and R.P. Lubianez, Macromolecules (1978) 11, 126.

(2) R.P. Lubianez, A.A. Jones and M. Bisceglia, Macromolecules (1980) 12, 1141.

(3) A.A. Jones and M. Bisceglia, Macromolecules (1979) 12, 1136.

(4) J.F. O'Gara, S.G. Desjardin and A.A. Jones, Macromolecules (1980) 14, 64.

(5) J.J. Connolly, E. Gordon and A.A. Jones, submitted to Marcomolecules.

(6) A.F. Yee and S.A. Smith, Macromolecules (1981) 14, 54.

(7) A.A. Jones and W.H. Stockmayer, J. Polym. Sci., Polym. Phys. Ed. (1977) 15, 847.

(8) T.A. Weber and E. Helfand, submitted to J. Chem. Phys.

(9)   A.S. Hay, F.J. Williams, G.M. Loucks, H.M. Relles, B.M.
      Boulette, P.E. Donabue and D.S. Johnson, Polym. Prepr. Am.
      Chem. Soc. Div. Polym. Chem. (1982) 23(2), 117.

(10) A.A. Jones and M.F. Tarpey, unpublished results.

(11) A.A. Jones, J. Polym. Sci., Polym. Phys. Ed. (1977) 15, 863.

(12) D.E. Woessner, J. Chem. Phys. (1962) 36, 1.

(13) W. Gronski and N. Murayama, Makrmol. Chem. (1978) 179, 1521.

(14) W. Gronski, Makromol. Chem. (1979) 180, 1119.

(15) E. Helfand, J. Chem. Phys. (1971) 54, 4651.

(16) E. Helfand, Z.R. Wasserman, and T.A. Weber, Macromolecules
      (1980) 13, 526.

(17) J. Skolnik and E. Helfand, J. Chem. Phys. (1980) 72, 5489.

(18) A.A. Jones, J.F. O'Gara, P.T. Inglefield, J.T. Bendler,
      A.F. Yee, and K.L. Ngai, Macromolecules (1983) 16, 658.

(19) H.W. Spiess, Colloid. Polym. Sci. (1983) 261, 193.

(20) P.T. Inglefield, R.M. Amici, J.F. O'Gara, C.-C. Hung and
      A.A. Jones, submitted to Macromolecules.

(21) A.E. Tonelli, Macromolecules (1972) 5, 558.

(22) A.E. Tonelli, Macromolecules (1973) 6, 503.

(23) I. Tvaroska and T. Bleha, J. Mol. Struct. (1975) 24, 249.

(24) G.A. Jeffrey and R. Taylor, J. Comp. Chem. 1 (1980) 99.

RECEIVED September 22, 1983

Characterization of Molecular Motion in Solid Polymers by Variable Temperature Magic Angle Spinning ^{13}C NMR

W. W. FLEMING, J. R. LYERLA, and C. S. YANNONI

IBM Research Laboratory, San Jose, CA 95193

The inclusion of a variable temperature magic-angle spinning capability for solid state ^{13}C NMR spectroscopy makes feasible the investigation by ^{13}C relaxation parameters of structural and motional features of polymers above and below Tg and in temperature regions of secondary relaxations. Herein, we report variable temperature (50K to 323K) spectral data on semicrystalline poly(propylene) and glassy PMMA. Illustrative of the data are the T_1 and $T_{1\rho}$ results for isotactic poly(propylene) over the temperature range 50K to 300K. All carbons in the repeat unit show minima in T_1 and $T_{1\rho}$ which reflect methyl group reorientation motion at the appropriate measuring frequencies (15 MHz and 57 kHz). The $T_{1\rho}$ data for CH and CH_2 carbons indicate the importance of spin-spin as well as spin-lattice pathways in their rotating frame relaxation over much of the temperature interval studied. An interesting spectral observation is the strong motional broadening of the methyl group in the temperature region of the $T_{1\rho}$ minimum. These and other facets of the poly(propylene) data as well as similar data for PMMA are discussed with respect to their implications for insight into polymer chain dynamics in the solid state.

One of the principal advantages of CPMAS experiments is that resolution in the solid state allows individual-carbon relaxation experiments to be performed. If a sufficient number of unique resonances exist, the results can be interpreted in terms of rigid-body and local motions (*e.g.*, methyl rotation, segmental modes in polymers, *etc.*) (1,2). This presents a distinct advantage over the more common proton relaxation measurements, in which efficient spin diffusion usually results in averaging of relaxation behavior over the ensemble of protons to yield a single relaxation time for all protons. This makes interpretation of the data in terms of unique motions difficult.

0097–6156/84/0247–0083$06.00/0
© 1984 American Chemical Society

Relaxation parameters of interest for the study of polymers include 1) ^{13}C and ^{1}H spin-lattice relaxation times (T_{1C} and T_{1H}), 2) the spin-spin relaxation time T_2, 3) the nuclear Overhauser enhancement (NOE), 4) the proton and carbon rotating-frame relaxation times ($T_{1\rho}^{C}$ and $T_{1\rho}^{H}$), 5) the C-H cross-relaxation time T_{CH}, and 6) the proton relaxation time in the dipolar state, T_{1D} (2). Not all of these parameters provide information in a direct manner; nonetheless, the inferred information is important in characterizing motional frequencies and amplitudes in solids. The measurement of data over a range of temperatures is fundamental to this characterization.

The initial studies of carbon relaxation in polymers have emphasized T_1 and $T_{1\rho}$ measurements, which provide information on molecular motions in the MHz and kHz frequency ranges, respectively. Schaefer and Stejskal have carried out the pioneering work in their investigations of glassy polymers (1). In particular, they stress the utility of $T_{1\rho}$ measurements for probing the dynamic heterogeneity of the glassy state and as a potential source of insight into the mechanical and other physical properties of polymers at the molecular level. Garroway and co-workers (3) reported the first variable-temperature (VT-MAS) $T_{1\rho}$ results in their study of epoxy resins, and together with VanderHart, (4) have detailed the complications in extracting information on molecular motion from $T_{1\rho}$ experiments.

In this paper, we report the first extensive sub-ambient VT-MAS ^{13}C T_1 and $T_{1\rho}$ data on macromolecules. The emphasis of the study was placed on isotactic poly(propylene)(PP) and atactic poly(methylmethacrylate)(PMMA) as they represent semi-crystalline and glassy polymers, respectively. Specifics of the investigation were directed to the issue of elucidating sidechain and backbone motions from the high frequency relaxation experiments.

Experimental

The ^{13}C data at 15.1 MHz were acquired on a modified Nicolet TT-14 NMR system. The features of this spectrometer and of the spinning assembly have been reported previously (5). Samples were machined into the shape of Andrew-type rotors and used directly for the various studies. Temperature variation was achieved by cooling or heating the helium gas used for driving the rotor. The temperature was controlled to $\pm 2°C$ with a home built temperature sensing and heater/feedback network. Spin-lattice relaxation times, T_1 were collected using a pulse sequence developed by Torchia (6) which allows cross-polarization enhancement of the signals. The $T_{1\rho}$ data were determined at 57 kHz using $T_{1\rho}$ methodology of Schaefer *et al.* (1). The PP examined was a 90% isotactic, 70% crystalline sample. The PMMA was an atactic commercial polymer.

Results and Discussion

Figure 1 shows the CPMAS ^{13}C spectra of PP as a function of temperature. The interesting feature is the progressive broadening of the methyl resonance

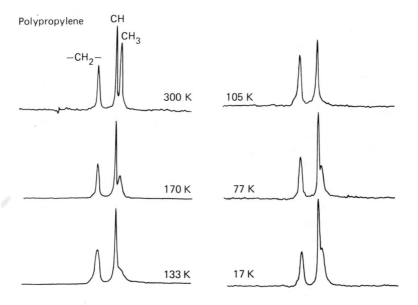

Figure 1. CPMAS ^{13}C spectra of poly(propylene) as a function of temperature.

as the temperature is lowered. At $\approx 110K$, the resonance is broadened to the point of disappearing from the spectrum. However, at temperatures below 77K, the methyl resonance narrows and reappears in the spectrum. This broadening phenomenon arises as the reorientation rate of the methyl group about the C_3 axis becomes insufficient to average (stochastically) dipolar interaction with the methyl protons. At the onset of broadening, the methyl motion has a correlation time comparable to the inverse of the strength (in frequency units) of the proton decoupling field. This reduces the efficiency of the rf decoupling and leads to a maximum linewidth of the carbon when the motions occur at the frequency corresponding to the amplitude of the proton decoupling field.

Rothwell and Waugh (7) have developed the theory for T_2 (the inverse of the ^{13}C linewidth) for an interplay between stochastic and coherent motions. For such a system, the profile of linewidth vs. temperature shows a maximum when the correlation time for molecular motion, τ_c, is equal to the modulation period of the decoupling, $(1/\omega_1)$. In the "short correlation time" limit $(\omega_1\tau_c<<1)$ (high temperature), the linewidth is reduced by the rapid motional averaging, while in the "long correlation time" limit $(\omega_1\tau_c>>1)$ (low temperature), the linewidth is reduced by efficient decoupling of C-H dipolar interactions. The spectra of PP in Figure 1 are consistent with the progression of the methyl resonance through the linewidth regions as the temperature is lowered. The reappearance (narrowing) of the methyl resonance at 77K indicates that the "long correlation time" regime has been reached (7). Further proof of the progressive changes in correlation time for methyl rotation as the temperature is lowered is provided by the decoupling-field dependence of the linewidth. At about 160K, the methyl linewidth is independent of decoupling field, while at 77K the linewidth varies as the inverse square of the decoupling field. This is the expected dependence for the transition between the extreme narrowing and long correlation time regimes (7). Finally, from the expression for the ^{13}C linewidth derived by Rothwell and Waugh [Eq. (1)] and the correlation time and temperature of the $T_{1\rho}^H$ minimum observed by McBrierty et al. (8),

$$\frac{1}{T_2} = \frac{\gamma_C^2\gamma_H^2\hbar^2}{5r_{CH}^6}\left(\frac{\tau_c}{1+\omega_1^2\tau_c^2}\right) \tag{1}$$

we calculate, for $\omega_1=57$ kHz (the value of the decoupling field used to obtain the spectra in Figure 1), that the maximum broadening for the methyl resonance would occur at 109K, in excellent agreement with our observations.

This broadening of the methyl resonance observed in PP is also found in polycarbonate, PMMA, and epoxy polymers. It should be a general phenomenon for rapidly reorienting side groups or main-chain carbons in polymers. For semicrystalline systems, where the local molecular structure is relatively homogeneous, severe broadening should result in the "disappearance" of resonance lines from the spectra. For glassy systems, where there is more heterogeneity in the local molecular environment, the effect may result in significant changes in resonance lineshape as a function of temperature as the carbons in differing environments undergo severe broadening. Of course, the phenomenon may be used to determine τ_c for the group undergoing the motion (7); however, the severe broadening does limit the ability to measure high-frequency relaxation times in such temperature intervals.

    The ^{13}C spin-lattice relaxation times for isotactic PP are shown in Figure 2. Primarily, the data represent that of the crystalline component. The semilog plots of intensity *vs.* time were nearly exponential for each of the carbons at all temperatures. Over the temperature range, each carbon in the repeat unit displays an individual relaxation time. The methyl relaxation appears to be dominated by methyl C_3 reorientation. If it is assumed that a C-H heteronuclear relaxation mechanism is operative, a calculation of the methyl carbon relaxation time based on a Bloembergen-Purcell-Pound (BPP) formalism and the correlation time at the proton T_1 minimum (8) at -110°C gives a value of 10 ms at -110°C, in good agreement with the observed value of 17 ms. In addition, the methyl motion also seems to dominate the backbone relaxation. This is evidenced by the shorter T_1 observed for the methine carbon relative to methylene (despite there being two direct C-H interactions for the methylene carbon). Apparently, backbone motions are characterized by such small amplitudes and low frequencies that contributions from the direct C-H interactions to spectral density in the MHz region of the frequency spectrum are minor relative to those from side groups. The $1/r^6$ distance dependence of dipolar relaxation thus accounts for both the long T_1 values of CH and CH_2 carbons (one to two orders of magnitude) relative to the methyl carbon and the shorter T_1 values for methine carbons relative to methylene carbon. The fact that the observed T_1 minimum for CH and CH_2 carbons is close to that reported for a proton T_1 minimum (at 30 MHz) (8) in PP that was assigned to methyl reorientation provides unequivocal support for the dominance of the T_1 relaxation by methyl protons.

    The $T_{1\rho}$ data (Figure 3) for the CH and CH_2 carbons also give an indication of methyl group rotational frequencies. As the temperature is lowered below 163K, the T_1 for these carbons increases and the $T_{1\rho}$ decreases by roughly an order of magnitude between 163K and 95K, suggesting that the contribution of methyl proton motion to MHz spectral density is decreasing, while increasing in the kHz regime. The CH and CH_2 $T_{1\rho}$ do not change greatly over the temperature interval from 163K to ambient, and, in contrast to the T_1 behavior, the CH_2 carbon has the shorter $T_{1\rho}$. The interpretation of the carbon $T_{1\rho}$ data is complicated by the fact that spin-spin (cross-relaxation) processes, as well as rotating-frame spin-lattice processes, contribute to the relaxation (4). Only the latter provide direct information on molecular motion. Although both processes show a dependence on the number of nearest-neighbor protons, the relative insensitivity of $T_{1\rho}$ to temperature and the approximate 2:1 ratio of CH_2/CH T_{CH} values also suggest that spin-spin processes dominate the relaxation above 163K. (If spin-spin effects dominate the rotating-frame relaxation and the carbon cross-relaxation to the proton

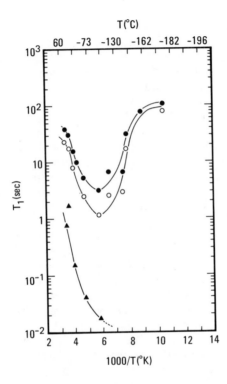

Figure 2. The ^{13}C spin-lattice relaxation times at 15 MHz for isotactic poly(propylene) methylene (●), methine (O), and methyl (▲) carbons.

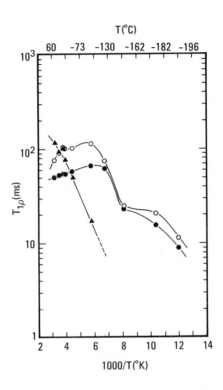

Figure 3. The $T_{1\rho}$ data at 57 kHz for CH (O), CH_2 (●), and CH_3 (▲) carbons in poly(propylene).

dipolar reservoir is less efficient than the corresponding proton spin-lattice relaxation T_{1D}, the observed $T_{1\rho}$ for the CH carbon will be about 1.7-2.0× that of the CH_2 carbon, based on the approximate twofold difference in the second moments (due to protons) for the two types of carbons. This is roughly the result observed in the data displayed in Figure 3.) Below 163K, the $T_{1\rho}$ of both carbons shorten and tend toward equality, indicating that spin-lattice processes derived from methyl reorientation are becoming competitive with the spin-spin process in relaxing the backbone carbon magnetization. McBrierty et al. (8) report a proton $T_{1\rho}$ minimum at 97K, which reflects methyl reorientation at kHz frequencies. No clear minimum is observed in the ^{13}C data, perhaps due to the interplay of the spin-spin and spin-lattice processes. Nonetheless, it is apparent that the methyl protons are responsible for the spin-lattice contributions to the CH and CH_2 $T_{1\rho}$ values.

Further evidence for the effect of spin-spin processes on $T_{1\rho}$ in PP is given in Figure 4, which shows $T_{1\rho}$ plotted against the reciprocal of the temperature as a function of the rotating frame field. As indicated in Table 1, in the case of motion dominating $T_{1\rho}$, there is a square dependence of $T_{1\rho}$ on field. For spin-spin domination, there is an exponential dependence. The results at room temperature clearly display a dependence greater than the 4× suggested for motion and the field variation. Only at temperatures less than 150K with large rotating-frame fields are strong motional effects observed. As previously discussed, these arise from methyl rotation.

The domination of both spin-lattice relaxation times for CH and CH_2 carbons in PP by methyl reorientation is clearly disappointing, since the potential for information on backbone motion due to the high resolution of the CPMAS experiments is not realized. The implication is that it may not be possible to observe backbone motion in crystalline materials having rapidly reorienting side groups without resorting to deuterium substitution of these side groups.

The T_1 data for various carbons in PMMA are given in Figure 5. Clear deviations from nonexponential behavior of the magnetization were often observed. Behavior different from that observed for PP presumably arises because the high degree of stereoregularity and high crystallinity of the PP provide a more homogeneous local environment than in glassy PMMA, where distributions of relaxation times are commonly observed, owing to site heterogeneity. For PMMA, the reported relaxation times represent the long-time portion of the magnetization decay curves.

The results for PMMA tend to cluster over the temperature range studied, except for the α-methyl carbon. The rapid relaxation for this carbon in the temperature range from 20°C to -70°C is consistent with the proton T_1 minimum at about -23°C assigned to α-methyl rotation at MHz frequencies (9).

The $T_{1\rho}$ data for PMMA are summarized in Table 2. As in the case of PP, the α-CH_3 undergoes motional broadening and disappears from the spectrum near the minimum in $T_{1\rho}$. In PMMA, severe broadening occurs in the temperature range between 140K and 200K. At lower temperatures, the

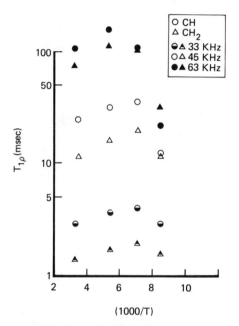

Figure 4. The $T_{1\rho}$ for methine (circles) and methylene (triangles) carbons in poly(propylene) as a function of the rotating-frame field and temperature; (a) 33 kHz, (b) 45 kHz, and (c) 63 kHz.

TABLE 1. ^{13}C $T_{1\rho}$ formalism.

Motion: BPP dipolar

$$\frac{1}{T_{1\rho}^C} = \frac{N_H\gamma_H^2\gamma_C^2}{20r_{CH}^6} f(\tau_c)$$

$$f(\tau_c) = \left\{ 2J(\omega_1) + \frac{J(\omega_H-\omega_C)}{2} + \frac{3J(\omega_c)}{2} + 3J(\omega_H) + 3J(\omega_H + \omega_c) \right\}$$

where

$$J(\omega_i) = 2\tau_c / \left(1 + \omega_i^2\tau_c^2\right) .$$

For long τ_c, $\dfrac{1}{T_{1\rho}^C} \propto \dfrac{1}{\omega_i^2}$.

Spin effects:

$$\frac{1}{T_{1\rho}^C} = \frac{1}{2} \sin^2\theta_s \, M_{CH}^{(2)} J_D(\omega_{1C})$$

$$J_D(\omega) = \pi\tau_D e^{-|\omega_{1C}|\tau_D}$$

$$\frac{1}{T_{1\rho}^C} \propto e^{2\pi\nu_{1C}\tau_D}$$

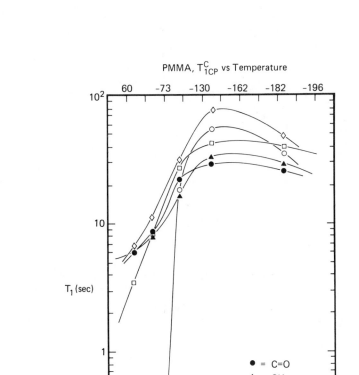

Figure 5. The T_1 data for various carbons in poly(methyl methacrylate).

TABLE 2. ^{13}C $T_{1\rho}$ data for poly(methyl methacrylate) (PMMA).

Temperature (K)	Carbon T_1(ms)			
	$-CH_3$	$-C=O$	$-C-$	$-OCH_3$
293	26	271	187	135
253	14	343	174	227
213	5	134	118	164
193	--	139	98	156
173	--	102	104	124
133	11	294	158	157
113	31	366	186	132
78	23	314	178	82

α-methyl carbon line narrows as the decoupling becomes effective. The $T_{1\rho}$ data for the quaternary carbonyl and methoxy carbons also reflect the methyl motion. Again, this demonstrates the effectiveness of methyl motion in relaxing other carbons in solids via long-range C-H dipolar interactions. Thus, despite the resolution of the CPMAS experiment, the extraction of information on local motional rates from carbon T_1 and $T_{1\rho}$ data on PMMA is not straightforward.

These initial results on the temperature dependence of ^{13}C spin-lattice relaxation times indicate the importance of VT-MAS experiments in the interpretation of relaxation pathways and suggests such experiments will help sort out the various motional processes.

Literature Cited

1. Schaefer, J.; Stejskal, E. O.; Buchdahl, R. Macromolecules 1975, 8, 291-296 and 1977, 10, 384-405.
2. Schaefer, J.; Sefcik, M. D.; Stejskal, E. O.; McKay, R. A. Macromolecules 1981, 14, 188-192 and references therein.
3. Garroway, A. N.; Moniz, W. B.; Resing, H. A. ACS Symposium Series 1979, 103, 67-87.
4. Garroway, A. N.; VanderHart, D. L. J. Chem. Phys. 1979, 71, 2772-2787.
5. Fyfe, C. A.; Mossbrugger, H.; Yannoni, C. A. J. Magn. Reson. 1979, 36, 61-68 and Lyerla, J. R. Contemp. Top. Polym. Sci. 1979, 3, 143-213.
6. Torchia, D. A. J. Magn. Reson. 1978, 30, 613-616.
7. Rothwell, W. P.; Waugh, J. S. J. Chem. Phys. 1981, 74, 2721-2732.
8. McBrierty, V. J.; Douglass, D. C.; Falcone, D. R. J. Chem. Soc., Faraday Trans. II 1972, 68, 1051-1059.
9. Powles, J. G.; Mansfield, P. Polymer 1962, 3, 336-339.

RECEIVED December 10, 1983

NEW TECHNIQUES IN NMR
OF LIQUID POLYMERS

New NMR Experiments in Liquids

GEORGE A. GRAY

Varian Associates, Palo Alto, CA 94303

The recent past has been the most exciting and pro-
ductive in the short history of NMR as an analytical
technique. The previous developments of homogeneous
and higher field magnets, multinuclear capabilites
and the introduction of pulsed and fourier methods
were spaced over a much longer time and new experiments took years to gain widespread attention and
use. New experiments were tied to new hardware and
generation of new approaches relied on performing
instrument development. Because of the "home-built"
and specialized nature of technique development,
rapid deployment of new experiments was impossible.

Particularly within the last five years, new
instruments have been introduced which have
accelerated the development of new experiments, and
at the same time made their widespread use possible.
The result has been an explosion in the number of
strategies and applicable approaches. This chapter
is intended to give an overview of these new
developments and to provide general classifications
by which advantages and drawbacks may be understood.
On modern instruments the instrument time which any
one experiment actually consumes is actually shrink-
ing, while the time necessary to intelligently
design and carry out the spectroscopic approach is
growing. Fortunately, the information gleaned from
the effort is growing at a much faster rate.

Instrumental Requirements. The topic "New Experiments" is specifically intended to cover new ways of
generating spectroscopic information. That is, the
techniques are not specifically oriented toward one

0097-6156/84/0247-0097$06.50/0
© 1984 American Chemical Society

nucleus or type of sample, but rely on fundamentally new methods of perturbing the nuclear spin system. The resulting signals are detected in the tradi- tional manner. Hence, the focus in "New Experi- ments" is on the events prior to signal acquisition. There is nothing unique about pulses of rf in this classification. CW methods can and have been imple- mented. However, the instrumental convenience and the ability to "instantaneously" affect nuclear spins gives high-power, pulsed NMR a decided advan- tage. Thus, the term "pulse sequence" has come to symbolize the recipe for "preparing" a nuclear spin system. Variations of placement, timing and phases of these pulses, on one or more nuclei, provide the richness and diversity upon which the rapid develop- ments of the last five years rest.

Spectral Editing Using J-Modulated Spin-Echos

The favorable properties of 180° refocussing pulses have been exploited in two main efforts; obtaining more spectral information by causing spectra to depend on J and the number of coupled nuclei, and secondly, discussed later, in polarization transfer methods.

Of course, the simplest way to make spectra depend on J, either homonuclear or heteronuclear, is to run a coupled spectrum. The low sensitivity of ^{13}C, along with the severe overlap in a coupled spectrum, make this approach not very universal. Frequently, an intermediate level of information is required, e.g., just a knowledge of the number of attached protons for the various carbons in a ^{13}C NMR spectrum. For many years the method of choice was the single-frequency off-resonance decoupling (SFORD) technique (1) which yields partially col- lapsed multiplets in the X spectrum for XH_n spin systems. The multiplets are broad due to uncol- lapsed long-range couplings and are difficult to interpret in complex molecules. The desirable features of broadband-decoupled spectra are compel- ling, particularly for small samples or congested spectra.

Fortunately, ^{13}C spin echo pulse sequences have been developed (2-17) which allow broadband- decoupling and carbon multiplicity selection (the number of attached protons). Patterned after an experiment introduced for studying proton-proton

couplings (18), the central theme is the amplitude
and phase modulation of, for example ^{13}C signals,
due to heteronuclear spin coupling. This modulation
describes the action of ^{13}C magnetization generated
by a pulse under the condition of coupling to
another spin -- therefore, the decoupler must be
turned off for some period of time. The effect may
be generated in either of two ways, gated (inter-
rupted) decoupling or by applying simultaneous proton and ^{13}C $180°$ pulses. The former pulse sequence
is noted in Figure 1 along with the effects of variation of τ within the pulse sequence. Three useful
characteristics of this approach are obvious: (1)
quaternary carbons are unmodulated, (2) at $\tau = 1/2J$
all protonated carbons are nulled, and (3) at $\tau =$
$1/J$ quaternaries and methylene carbons are at opposite phase from methyl and methine carbons.

The investigator thus has the option of search-
ing for quaternaries which might be masked by protonated carbons by setting $\tau = 1/2J$; or, it might be
more desirable to sort by CH/CH_3 and CH_2/C. It is
also possible to combine normal spin-echo and modulated spin-echo spectra to arrive at edited subspectra containing only carbons of one type (6,14-17).

Patt and Shoolery (8) have addressed the problem of optimizing sensitivity in this experiment in
their APT (Attached Proton Test) experiment. The
normal use of a $90°$ pulse to begin the sequence has
the effect of saturating longer T_1 nuclei and therefore requires longer equilibrium delays. In normal
experiments it is customary to use an intermediate
pulse angle set relative to the expected T_1's and
repetition rate of the experiment. By incorporating
a second $180°$ pulse just prior to acquisition they
show that z-magnetization left after an initial
sub-$90°$ pulse is restored to the z-axis by the last
$180°$ pulse, thereby allowing faster repetition of
the experiment.

An example of the use of the APT is given in
Figure 1 for an ethylene-1-hexene copolymer.
Inverted signals arise from CH and CH_3 carbons.
Contrast this direct determination to the more
poorly-resolved and lower sensitivity off-resonance
data also given.

The delayed detection aspects of the spin-echo
technique can be used to eliminate protonated

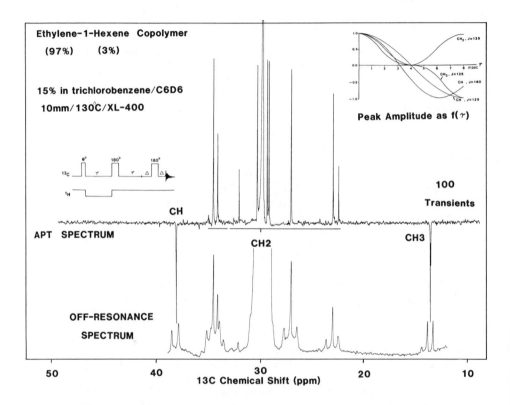

Figure 1. 100 MHz ^{13}C spectra of ethylene-1-hexene. The APT spectrum was obtained with a tau of 7 milliseconds, inverting the methylene carbons. Figure courtesy of Varian Associates.

carbons from a ^{13}C spectrum. However, no J-modulation period need be used. Rather, the proton decoupler is placed off-resonance and is either CW or broadband-modulated. The distance off-resonance must be enough to severely broaden the resonances of protonated carbons, without significantly affecting the linewidth (or T_2) of nonprotonated carbons. By delaying acquisition for a short period following an initial 90° pulse the signals for protonated carbons decay rapidly and are not detected. The quaternary carbons are detected easily and phasing the spectrum is made simpler by inserting a refocussing 180° pulse midway between the 90° pulse and acquisition. Cookson and Smith (14) have applied this technique to petroleum mixtures, clearly detecting the quaternary aromatic carbons. Extension of this method to non-protonated carbons in polymers is direct and straightforward, allowing direct detection of branching sites, for example, which are hidden underneath a large protonated carbon envelope.

Beloeil, et. al. (12) have introduced a new type of spin-echo sequence which produces either C/CH_2 or CH/CH_3 ^{13}C subspectra. It involves an initial 45° pulse followed by τ-180°-τ period and a final monitoring 45° pulse either in phase or 180° out of phase with respect to the original 45° pulse. The proton decoupler is gated off during the second $\tau(= 1/J)$ period.

Polarization Transfer Methods

Those familiar with the routine acquisition of ^{13}C NMR spectra are aware of the consequences of the nuclear Overhauser effect (NOE). Saturation of protons has the effect of increasing the net ^{13}C magnetization of those carbons relaxed by the protons of up to a factor of three times the equilibrium magnetization. Most analytical or survey ^{13}C spectra are obtained with continuous broadband proton decoupling and any resultant NOE. Characteristics of this mode of operation are, (1) the possibility of variable NOE, (2) repetition rate governed by ^{13}C T_1 and (3) both protonated and non-protonated carbons are detected. The first aspect makes quantitation difficult. The second affects net sensitivity, and the third has the prospect of having undesirable signals in certain situations.

Since quantitation is crucial in many polymer analyses, it is important to obtain data with T_1 and NOE in mind. Highly flexible polymers undergoing rapid segmental motion typically give narrow ^{13}C lines. Often these carbons can have T_1's of several seconds and full or nearly full NOE. Other, more rigid polymers may exhibit broad lines and little NOE. In the case of ethylene-1-hexene copolymer there is considerable NOE for the 1-hexene portion (Figure 2). Relative peak areas can produce good concentrations only if T_1 and NOE are properly considered.

A fundamentally different approach to signal excitation is present in polarization transfer methods. These rely on the existence of a resolvable J coupling between two nuclei, one of which (normally the proton) serves as a polarization source for the other. The earliest of these type of experiments were the SPI (Selective Population Inversion) type (19) in which low-power selective pulses are applied to a specific X-satellite in the proton spectrum for an X-H system. The resultant population inversion produces an enhanced multiplet in the X spectrum if detection follows the inversion. A basic improvement which removes the need for selective positioning of the proton frequency was the introduction of the INEPT (Insensitive Nucleus Excitation by Polarization Transfer) technique by Morris and Freeman (20). This technique uses strong non-selective pulses and gives general sensitivity enhancement.

The pulse sequence has a basic polarization transfer portion (Figure 3) which produces a net inversion of one of the proton spin states. Following an X nucleus 90° pulse there exists enhanced magnetization in the X multiplet. The signal enhancement is proportional to the ratio of the magnetogyric ratios of the two nuclei involved, a factor of 4 for ^{13}C and 10 for ^{15}N for X-^1H experiments. The repetition rate of the experiment is dictated by the T_1 of the polarization source nucleus. Since this is typically the proton, more signal per unit time often can be obtained than by NOE. Since the relaxation mechanism for the X nucleus is not involved, variable or negative NOE is not a problem, as can be the case for ^{15}N and ^{29}Si. One drawback is the effect of short T_2's on the sensitivity improvement. Since the sequence requires

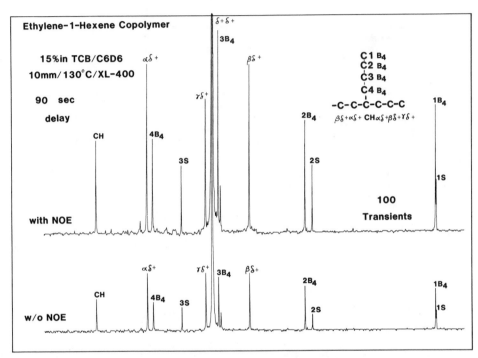

Figure 2.   Comparison of ^{13}C spectra with and without nuclear overhauser enhancement. Figure courtesy of Varian Associates.

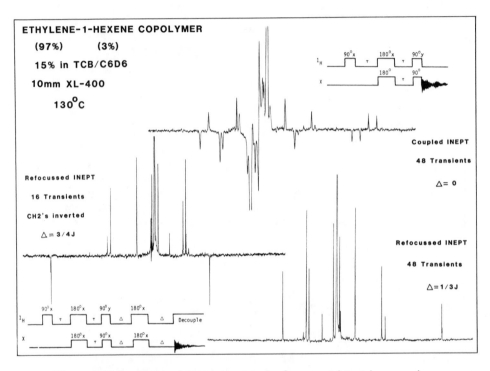

Figure 3.   Polarization transfer excitation using
INEPT. Figure courtesy of Varian Associates.

delays on the order of 1/J before detection, (X-H coupled systems with J of 10-100 Hz require delays of from 100-10 ms) short proton T_2's or X nucleus T_2's in decoupled polarization transfer experiments can radically lower expected gains. For example, if the protons have a T_2 of 30 ms (linewidth in the proton spectra of ∿10 Hz), 2/3 of the proton magnetization will decay away in 30 ms. This would be the same time necessary to attain polarization transfer for a system having an X-H J coupling of 16 Hz. For typical one-bond couplings of 40-250 Hz for ^{15}N and ^{13}C, this loss is less important because of the 1/J delay dependence. Although NOE falls off as field strength and molecular size increases, it still may have a net advantage for macromolecules because of T_2.

While the first reported INEPT experiment dwelt on a generation of enhanced coupled spectra, later reports (21-23) extended the basic sequence to allow broadband decoupling during detection, realizing the goal of general sensitivity enhancement in a more widely useful form. As a side benefit, through proper selection of a refocussing period, spectral editing is possible. No quaternary carbons are usually observed since delays are typically set for J's appropriate for protonated X nuclei. XH only spectra can be obtained easily, as well as spectra in which XH_2 is inverted with respect to XH and XH_3. The combination of these two spectra then allows direct identification of XH, XH_2 and XH_3 signals in even the most complex molecules in a similar manner as in the J modulated spin echo experiments described above.

An analogous case is present in the ^{13}C-^2H INEPT experiment (24). Here, only deuterated carbons are observed. While the magnetogyric ratio factor should give only a theoretical 3/5 enhancement, the short ^2H T_1 allows very rapid accumulation. The technique should be useful for determination of deuterated X nuclei spectra without interferences from protonated X signals. The sensitivity gain should be very dramatic for fully deuterated X nuclei where normal X nucleus T_1's can be extremely long because of inefficient dipolar relaxation from ^2H or other mechanisms. This experiment offers the capability of detection of deuterated sites at small concentration in, for example,

studies of the mechanism of deuteration (or reaction
using deuterated reactants, e.g., oxidants) of
macromolecules.

While possessing several advantages, INEPT has
properties which are undesirable: (1) coupled spectra have, at times, missing lines and antiphase components; (2) there is a fairly strong J dependence
in cases where signals should null, leading to resi-
dual unwanted signals; and, (3) refocussed INEPT
delay periods are variable, depending on multiplicity desired, leading to possible variation in
intensity due to variations in T_2 among the resonances. Doddrell, Pegg and Bendall (25,26) have
developed a Distortionless Enhancement by Polarization Transfer (DEPT) pulse sequence (Figure 4) which
addresses these problems to a large degree. Coupled
X nuclei DEPT spectra have the normal binomial distribution of intensities, the J dependence for XH
selectivity is better than for INEPT, and the multiplicity selection relies on a variable proton flip
angle rather than variable delays, thus factoring
out T_2 dependence. By choosing $\theta = 45^\circ$, 90°, and
135° in separate spectra, spectral editing is possible through proper combinations of these spectra to
give XH, XH_2 and XH_3 subspectra. DEPT is more sensitive overall to spin relaxation during the pulse
train since the relevant delay is of magnitude $3/2J$
while corresponding periods in INEPT ranges from $1/J$
to $5/4J$ typically.

Two-Dimensional NMR

Probably the greatest recent change in the practice
of NMR has been the explosive growth in techniques
and applications of 2D NMR.

2D NMR is an extension of ordinary NMR. The
basic principle was invented by Jenner (27) and covers essentially all 2D experiments in NMR, as well
as other areas. The general 2D NMR experiment is
characterized by up to four time periods:

Preparation...Evolution(t_1)...
Mixing(t_m)...Detection(t_2)

Preparation time is necessary to bring the system to
a known state, e.g., equilibrium magnetization, and
is usually a fixed delay time. At the beginning of
the Evolution period, the spins are perturbed and

DEPT Pulse Sequence

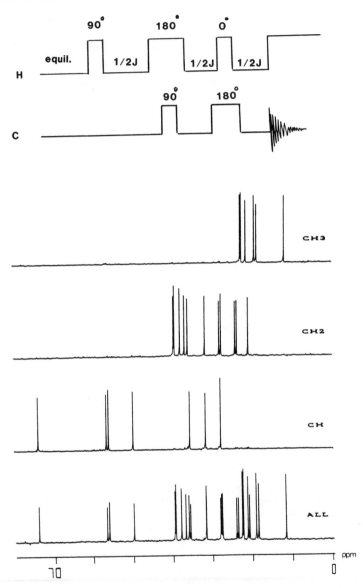

Figure 4. The edited spectra are the result of specific combinations of different spectra obtained for θ = 45°, 90° and 135° using a solution of cholesteryl acetate in CDCl₃. Figure courtesy of Varian Associates.

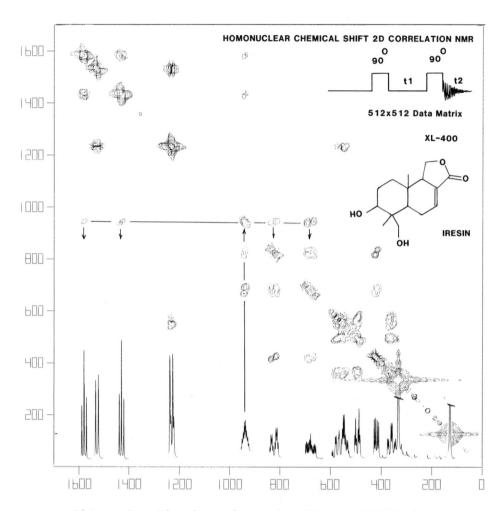

Figure 5. The data from the 2D experiment is
displayed in contours of intensity, each axis
represents a proton shift axis. The off-diagonal
intensities show homonuclear couplings. For
instance, the proton at 950 Hz (as indicated by
an arrow) has off-diagonal intensity showing
couplings to the four other indicated protons.
Figure courtesy of Varian Associates.

allowed to evolve within a specific environment that
may be different from any environment in the other
time periods. This Evolution time is regularly
varied through a range of values from zero to some
maximum, each resulting in a separate FID recorded
for a fixed time during the detection period (Certain 2D experiments use a mixing, or spin exchange
period).

The result is a collection of FID's, the
number of which equals the number of different
values of the Evolution time period. All FID's are
then transformed in the usual way, resulting in a
collection of spectra (the familiar inversion-
recovery T_1 experiment is an example of this stage
of the process). A new collection of FID's is
assembled by taking data points from each spectrum
at given frequency values.

For example, values of spectral intensity in
each spectrum found at a frequency f are taken and
arranged in a data table. This is repeated for
every f value in the spectrum, resulting in N FID's
when N is the number of points in the original spectrum. These N FID's are then fourier transformed,
resulting in N spectra. These N spectra, presented
on a two-dimensional plot, are now characterized by
two frequency axes. One frequency (F2) is always
that of the normal observe spectrum since it results
from FT of normal FID's. The significance of the
remaining (F1) frequency is determined by the pulse
sequence used.

There have been major advances in the last few
years both in the number and variety of the pulse
sequences used to perturb the spins, and in the com-
puting power and data processing techniques
employed.

All two dimensional experiments use essentially
the same data reduction process. Several years ago
the data acquisition phase of 2D experiments was
usually the shortest, with double transformation
requiring up to several hours followed by, again,
comparable plotting time. Off-line data processing
was often used for faster processing, but this
severely limited any widespread use of 2D methods.
2D processing capabilities became generally available on commercial spectrometers after 1979 and with
the introduction of flexible pulse programmers the

use of 2D methods has grown quickly. Very recently,
new computer hardware, including Array Processors,
has revolutionized 2D data processing, allowing complete data reduction, including full 2D display in
less than 30 seconds (for a typical 512 x 512
matrix). This speed, coupled with responsive color
graphics, has permitted direct, on-line data processing and removed the formidable time-barrier to
full use of 2D NMR spectroscopy.

J-Resolved 2D NMR. The key to the nature of the F_1
domain lies in the environment in which the spins
precess during the evolution period t_1. This
environment may include the interactions of spin
coupling, the chemical shift -- which governs the
rate of precession (relative to the transmitter frequency as zero) -- and magnet inhomogeneity which
causes identical nuclei to have different precession
rates. Refocussing pulses can remove some, or all,
of these interactions. In particular, after a $90°$
pulse use of a strong non-selective $180°$ refocussing
pulse on the observe nucleus midway through the evolution period eliminates the effect of chemical
shift differences for all observe spins as detected
in t_2 for any t_1. Hence, the resultant 2D spectrum
might be thought to be rather simple, as it is
indeed for non-coupled nuclei -- singlets are centered at coordinates ($F_1 = 0$, $F_2 = \nu_H$). However,
the mutual J coupling of spins during t_1 does affect
the phase of the signal when detected in t_2, and
hence J information is encoded in the detected signals.

    The heteronuclear J-resolved 2D experiment pro-
duces completely separate information in both
domains, X nucleus chemical shift in F_2 (since
broadband decoupling in t_2 removes the effect of J
coupling) and AX coupling in F_1. This is particularly valuable, for example, in $^{13}C-^1H$ cases since
complete coupling patterns can be extracted for each
^{13}C without overlap of adjacent patterns. Information can be extracted easily, in contrast to direct
observation of highly overlapped coupled 1D spectra.

    The homonuclear J-resolved 2D spectrum is dif-
ferent in that AX coupling is active in both t_1 and
t_2, and thus is reflected in the existence of spin
multiplets in both dimensions.

Each spin pattern is present although no one slice perpendicular to F_2 contains a whole pattern. A computer process known as "tilting", "rotating", or "shearing" can be used to realign the spin patterns ($45°$ tilt) perpendicular to F_2. This permits display and plotting of "slices" of the 2D data -- an F_1 spectrum at some specified F_2 value. These slices are the fully coupled spin patterns. The spin-echo nature of the experiment produces narrow lines, and thus very highly resolved multiplets. In highly congested, but first-order, proton spectra this technique can be of immense value.

Polarization transfer may be used in heteronuclear J-resolved 2D NMR (28) for signal enhancement, shorter equilibration periods and elimination of non-protonated resonances -- all the same features as in INEPT or DEPT. In this experiment the refocussing period following polarization transfer becomes the evolution time, rather than the $1/4J$, $1/2J$, and $3/4J$ values normally used for refocussed INEPT or variable flip angle θ for DEPT. The polarization transfer part of the sequence includes a $1/2J$ period, as usual, and thus the 2D sequence can be tailored to specific types (or J's) of XH coupled pairs, e.g., certain long-range couplings for quaternary carbons whose normal ^{13}C T_1's may be too long to permit ordinary heteronuclear 2D J experiments.

Chemical Shift Correlation 2D NMR. Probably the most widely used 2D NMR experiments are those which relate chemical shifts of different nuclei. That is, F_1 and F_2 represent chemical shift axes and the class includes both homonuclear and heteronuclear categories. Again, the environment during the evolution time dictates the interpretation of the new information. The simplest to consider is the homonuclear chemical shift correlation (27,29-31), also known in one form as COSY (correlation spectroscopy). The most widely used pulse sequence is rather simple, consisting of $90°-t_1-90°$ followed by acquisition. The resulting peaks in the 2D plot fall in several categories: axial, diagonal and off-diagonal. The axial peaks lie along $F_1 = 0$ and result from longitudinal magnetization sampled by the last pulse. The diagonal peaks are due to magnetizations which remain associated with the same spins before and after the second $90°$ pulse, that is the magnetization has the same frequency during t_1

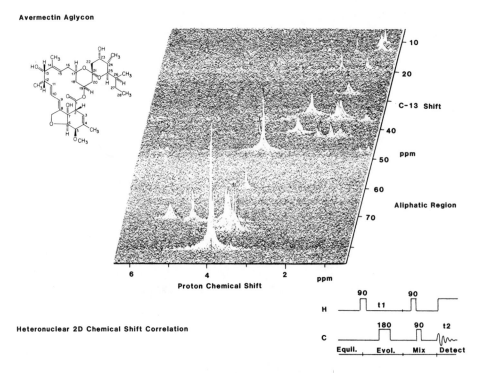

Avermectin Aglycon

Heteronuclear 2D Chemical Shift Correlation

Figure 6. Each peak in the 2D spectrum arises from a C-H bond. Extrapolation to the relevant axes give the corresponding chemical shifts. The data was obtained on 90 mg in 10.5 hours using a 10 mm probe at 75 MHz (XL-300). Figure courtesy of Varian Associates.

and t_2. The off-diagonal peaks confirm the fact that magnetization present at one frequency in t_1 is transferred to another frequency after the second 90° "mixing" pulse. This is only possible if the two nuclei at the two frequencies share the same spin system, that is, they are J-coupled. This last feature forms the basis of the widespread use of homonuclear shift correlated 2D NMR since it gives equivalent information to spin-decoupling. The spread of information in two dimensions actually makes the interpretation easier. Proper phase cycling (31) can remove the axial peaks, considerably simplifying the experiment. Variations on the basic experiment include the 90°-t_1-45°-acquire experiment (29). This version narrows the patterns on the diagonal permitting analysis of correlations of nuclei closely spaced in chemical shift.

The second major form of shift correlation 2D NMR is heteronuclear 2D shift correlation. The experiment as proposed by Maudsley and Ernst (32) can be visualized as generating proton magnetization, for example, by a 90° pulse and letting the magnetization precess for a time t_1. The extent of precession (in the rotating frame) is proportional to the distance off-resonance from the proton transmitter frequency and therefore its phase is proton shift dependent. Simultaneous 90° pulses on the proton and X observe nucleus transfer magnetization to X -- just as in the INEPT experiment. The phase of the X nucleus magnetization is hence coded with the chemical shift of the attached proton. The experiment is carried out for the full range of t_1 values to produce the 2D spectrum. Since the X nucleus shift is present in F_2, the XH pair has both chemical shifts identified by extrapolation to the individual axes from the XH peak in the 2D data.

Broader utilization of heteronuclear chemical shift 2D has really just begun. Amman et. al., (33) have illustrated how 2D methods can be used to assign protons and carbons in lupane, a C_{30} triterpene containing only carbon and hydrogen. Heteronuclear J-resolved 2D was used to assign the number of protons to each carbon, and heteronuclear chemical shift correlation allowed assignment of the associated proton shifts and many of the homonuclear H--H coupling constants. Ikura and Hikichi (34) used the same techniques on d-biotin. A mixture of allylnickel complexes was also studied by Benn (34), and

Morris and Hall (35) have examined a series of car-
bohydrates. Larger molecules are, of course, feasible. Chan and Markley (36) have assigned histidine,
tyrosine and phenylalanine protonated carbon resonances in uniformly enriched (20%) oxidized ferro-
dixin by heteronuclear H/C 2D shift correlation NMR.
Polypeptide resonances have also been assigned using
C/H 2D NMR (37,38). In these systems H/H homonuclear, H/C heteronuclear and N/H heteronuclear 2D
techniques have allowed direct confirmation of
assignments and connectivities, in particular the NH
nitrogens and protons.

Heteronuclear chemical shift correlation techniques can be used to infer spin-lattice relaxation
times of the protons attached to the observe nucleus
(39). This is accomplished by saturation of the protons and observe nucleus followed by a variable time
t (saturation recovery) during which the observe
nucleus is continued at saturation (by repeated
pulsing) and the attached proton remagnetizes. This
process is followed by the normal H/X 2D shift
correlation experiment. The t-dependence of the 2D
peak intensities is then used to extract ^{1}H T_1's by
exponential analysis. This approach can be used to
extract proton T_1's in polymers via observation of
their attached ^{13}C's, obtaining motional data normally impossible to obtain by proton NMR.

All of the above heteronuclear shift correlation techniques produce spectra which have H-H spin
multiplets in the F_1 dimension. These can be
invaluable in certain situations where the pattern
is obscured in the 1D spectrum. At other times sensitivity considerations would argue for collapsing
these multiplets, thereby gaining at least a factor
of two in intensity. In other cases, highly congested 2D spectra could have overlapped H/C correlations. Bax (40) has developed a method for collapsing these multiplets which essentially replaces the
single X nucleus 180° pulse at the mid-point of the
evolution time with the element $90°_x(H)-1/2J-$
$180°_x(H)180°_x(X)-1/2J-90°_x(H)$. The resultant 2D
shift correlation spectra are characterized by single peaks for a X-H bond. Slices in F_1 show proton
singlets at the appropriate chemical shift. Of
course, a projection onto the F_1 axis would give a
proton "stick" spectrum.

Magnetization Transfer 2D NMR. One of the more
exciting aspects of 2D NMR is the class of experiments which can probe intra-molecular interactions,
proximity of nuclei and chemical exchange. As
opposed to some of the above 2D experiments where
transverse (X,Y) magnetization transfer occurs,
there are experiments wherein longitudinal (Z) magnetization is transferred, i.e., energy level populations change. The nuclear Overhauser effect is
one example of population redistribution within a
spin system via an incoherent process. Other types
include saturation transfer experiments where the
mechanism is chemical exchange.

Other means of studying chemical exchange or
NOE have included lineshape analysis at or near
coalescence and selective saturation or inversion
with subsequent following of propagation of magnetization throughout the spin system. The former technique can be very difficult for broad coalescence
peaks and require certain temperatures to be established. The latter is very useful for a limited
number of lines to be affected but gets time-
consuming for more than a few lines and very difficult for closely-spaced lines. The 2D method
requires no particular special conditions and therefore has a great deal of attractiveness for studies
of chemical and biochemical dynamics. This class of
experiments was proposed by Jenner et. al. (41,42)
and elaborated on by Macura and Ernst (43). The
basic pulse sequence is 90°-t_1-90°-$t(MIX)$-90°. The
second 90° pulse can be thought of as restoring to
the Z-axis transverse magnetizations created by the
first 90° pulse. The component and direction of the
resultant Z-axis magnetization is a function of the
precession of the spins during t_1. Some components
will be positive, some negative and some zero.
These components will oscillate as a function of t_1.
The resultant Z-magnetizations will be in a non-
equilibrium state and spin-lattice relaxation
processes will work toward equilibrium. For the
case of protons relaxation is primarily dipolar via
other protons, this non-equilibrium magnetization
will then be redistributed by mutual spin flip to
other protons. The final 90° pulse monitors the
extent of magnetization transfer. The 2D experiment
samples all degrees of magnetization transfer as it
increments t_1.

The 2D shift correlation spectrum thus produced
is characterized by the usual diagonal peaks coming
from magnetization remaining at the same frequency
in t_1 and t_2. Phase-cycling can remove the axial
peaks, just as in homonuclear shift correlation 2D
NMR. The off-diagonal peaks come from magnetization
which has been transferred in a spin-lattice relaxation sense from one type of spin to another. This
could be because of true chemical exchange occurring
during the mix time, or it could be simply from
spins close enough in space to provide mutual dipolar relaxation. This sequence has been used to
establish NOE's in polypeptides and proteins
(45,46). It is particularly effective in macromolecules where there is much slower molecular tumbling and more favorable NOE's. Variation of t(MIX)
allows probing of the rates of magnetization
transfer and thus proximity of protons for NOE, or
chemical kinetics for true exchange (47).

Carbon-13 chemical exchange networks have been
explored by Huang, Macura and Ernst (48) using the
above techniques. Normal $90^o-t_1-90^o-t(MIX)-90^o$
sequences were used, in the presence of proton
decoupling (for NOE, sensitivity and simplicity).
Refocussed INEPT was also used to prepare the initial magnetization, in place of the first 90° pulse,
to gain additional sensitivity. This extension suggests future flexibility and selectivity since the
INEPT portion of the sequence can be tailored for
specific J's (long-range and direct), thus allowing
very precise control over the site from which magnetization can evolve. Huang, et. al (48) applied
these strategies to the classic exchange problems of
ring-puckering in decalin, bond shift in bullvalene
and solvation shell exchange in aluminum complexes.
The longer relaxation times of carbon-13 can actually be put to advantage in these studies since they
permit longer mix times for slower exchange
processes. The other major advantage of the 2D
technique is that it can be performed on a very
slowly exchanging system whose lines are still narrow.

Literature Cited

1. Ernst, R.R., J. Chem. Phys. 1966, 45, 3845;
   Reich, H.J.; Jantelat, M.; Meese, M.T.;
   Weigert, F.J.; J.D. Roberts, J. Am. Chem. Soc.
   1969, 91, 7445.
2. Anet, F.A.L.; Jaffer, N.; Strouse, J., 21st
   Experimental NMR Conference, 1980, Tallahassee,
   FL.
3. Rabenstein, D.L.; Nakashima, T.K., Anal. Chem.
   1979, 51, 1465A.
4. Levitt, M.H.; Freeman, R., J. Magn. Reson. 1980,
   39, 533.
5. LeCocq, D.; Lallemand, J.-Y., J. Chem. Soc.
   Chem. Commun. 1981, 150.
6. Cookson, D.J.; Smith, B.E., Org. Magn. Reson.
   1981, 16, 11.
7. Brown, D.W.; Nakashima, T.K.; Rabenstein, D.L,
   J. Magn. Reson. 1981, 45, 302
8. Patt, S.L.; Shoolery, J.N., J. Magn. Reson. 1982,
   46, 535.
9. Cookson, D.J.; Smith, B.E., Fuel 1983, 62, 34.
10. Pei, Feng-Kui, Freeman, R., J. Magn. Reson.
    1982, 48, 318.
11. Jakobsen, H.J.; Sorensen, O.W.; Brey, W.S.;
    Kanyha, P., J. Magn. Reson. 1982, 48, 328.
12. Beloeil, J.-C.; LeCocq, C.; Lallemand, J.-Y.,
    Org. Magn. Reson. 1982, 19, 112.
13. Nakashima, T.K.; Rabenstein, D.L., J. Magn.
    Reson. 1982, 47, 339.
14. Cookson, D.J.; Smith, B.E., Fuel 1983, 62, 39.
15. Bendall, M.R.; Doddrell, D.M.; Pegg, D.T., J.
    Amer. Chem. Soc. 1981, 103, 4603.
16. Bendall, M.R.; Pegg, D.T.; Doddrell, D.M.;
    Johns, S.R.; Willing, R., J.C.S. Chem. Commun.,
    1982, 1138.
17. Bendall, M.R.; Pegg, D.T.; Doddrell, D.M.;
    Williams, D.H., J. Org. Chem. 1982, 47, 3021.
18. Campbell, I.D.; Dobson, C.M.; Williams, R.J.P;
    Wright, P.E., FEBS Lett. 1975, 57, 96.
19. Pachler, K.G.R.; Wessels, P.L., J. Magn. Reson.
    1977, 28, 53; Jakobsen, H.J.; W.S. Brey, J.
    Amer. Chem. Soc. 1979, 101, 760.
20. Morris, G.A., J. Amer. Chem. Soc. 1980, 102, 428.
21. Burum, D.P.; Ernst, R.R., J. Magn. Reson. 1980
    39, 163.
22. Doddrell, D.M.; Pegg, D.T., J. Amer. Chem. Soc.
    1980, 102, 6388.
23. Bolton, P.H., J. Magn. Reson. 1980, 41, 287.

24. Rinaldi, P.L.; Baldwin, N.J., J. Amer. Chem.
    Soc. 1982, 104, 5791.
25. Bendall, M.R.; Pegg, D.T.; Doddrell, D.M.;
    Field, J., J. Amer. Chem. Soc. 1981, 103, 934.
26. Dodrell, D.M.; Pegg, D.T.; Bendall, M.R., J.
    Magn. Reson. 1982, 48, 323.
27. Jeener, J.; Ampere International Summer School,
    Basko Polje, Yugoslavia 1971.
28. Thomas, D.M.; Bendall, M.R.; Pegg, D.T.;
    Doddrell, D.M.; Field, J., J. Magn. Reson. 1981,
    42, 298, Rutar, V.; Wong, T.C., J. Magn. Reson.
    1983, 53, 495.
29. Aue, W.P.; Bartholdi, E.; Ernst, R.R., J. Chem.
    Phys. 1976, 64, 2229.
30. Nagayama, N.; Kumar, A.; Wuthrich, K.; Ernst,
    R.R., J. Magn. Reson. 1980, 40, 321.
31. Bax, A.; Freeman, R.; Morris, G., J. Magn.
    Reson. 1981, 42, 164.
32. Maudsley, A.A.; Ernst, R.R., Chem. Phys. Lett.
    1971, 50, 368.
33. Ammann, W.; Richarz, R.; Wirthlin, T.; Wendisch,
    D., Org. Magn. Reson. 1982, 20, 260.
34. Ikura, M.; Hikichi, K., Org. Magn. Reson. 1982,
    20, 266.
35. Benn, R., Z. Naturforsch 1982, 37b, 1054.
36. Morris, G.A.; Hall, L.D., J. Amer. Chem. Soc.
    1981, 103, 4703.
37. Chan, T.-M.; Markley, J.L., J. Amer. Chem. Soc.
    1982, 104, 4010.
38. Kessler, H.; Hehlein, W.; Schuck, R., J. Amer.
    Chem. Soc. 1982, 104, 4534.
39. Gray, G.A., Org. Magn. Reson. 1983, 21, 111.
40. Avent, A.G.; Freeman, F., J. Magn. Reson. 1980,
    39, 169.
41. Bax, A., J. Magn. Reson. 1983, 53, 517.
42. Jeener, J.; Meier, B.H.; Bachmann, P.; Ernst,
    R.R., J. Chem. Phys. 1979, 71, 4546.
43. Meier, B.H.; Ernst, R.R.; J. Amer. Chem. Soc.
    1979, 101, 6441.
44. Macura, S.; Ernst, R.R., Mol. Phys. 1980, 41,
    95.
45. Kumar, A.; Ernst, R.R.; Wutrich, K., Biochem.
    Biophys. Res. Commun. 1980, 95, 1.
46. Bosch, C.; Kumar, A.; Baumann, R.; Ernst, R.R.;
    Wutrich, K., J. Magn. Reson. 1981, 42, 159.
47. Kumar, A.; Wagner, G.; Ernst, R.R.; Wutrich, K.,
    J. Amer. Chem. Soc. 1981, 103, 3654.
48. Huang, Y.; Macura, S.; Ernst, R.R., J. Amer.
    Chem. Soc. 1981, 103, 5327.

RECEIVED October 31, 1983

Application of the INEPT Method to ^{13}C NMR Spectral Assignments in Low-Density Polyethylene and Ethylene–Propylene Copolymer

KUNIO HIKICHI, TOSHIFUMI HIRAOKI, and SHINJI TAKEMURA—Department of Polymer Science, Hokkaido University, Sapporo 060, Japan

MUNEKI OHUCHI—NMR Application Laboratory, Scientific Instrument Division, JEOL Ltd., Akishima, Tokyo 196, Japan

ATSUO NISHIOKA—Faculty of Engineering, Fukui Institute of Technology, Gakuen, Fukui 910, Japan

Carbon 13 NMR spectra of low density-polyethylene and ethylene propylene copolymer were observed at a freqeuncy of 125.77 MHz. The INEPT method was applied to identify methyl, methylene, and methine carbons. It was demonstrated that the INEPT method provides a very useful technique for characterization of polymers.

It has been well established that ^{13}C NMR spectroscopy provides an important technique for characterization of various polymers(1-3). Short-chain branches in low-density polyethylenes and monomer sequence distributions in ethylene propylene copolymers have been the subject of a number of studies(4-27), because the type, distribution, and concentration of branches in a low-density polyethylene and the sequence distribution in a copolymer should greatly affect the thermodynamic, morphological, and physical properties of the polymers. Although many ^{13}C NMR studies have been reported on low-density polyethylenes and ethylene propylene copolymers, the results are still ambiguous. The reason for this is the difficulty in obtaining complete assignments for each resonance in terms of contributing branches or sequences. So far the peak assignment has been made through various means: a comparison of the observed chemical shift values with those calculated by empirical additivity rules(28-30), a consideration of internal consistency of intensities, a comparison of alkane model compounds, and specific labelling.

Four years ago, Morris and Freeman(31) reported a method for enhancing ^{13}C NMR signals by using polarization transfer from protons(INEPT). Subsequently, Doddrell and Pegg(32) showed that this INEPT method could be used to identify methine, methylene, and methyl groups of a complex molecule when combined with appropriate delay time (Δ) prior to data acquisition and broad-band decoupling. If $\Delta=1/2J$ (J is the coupling constant between ^{13}C and ^{1}H), only methine carbon resonances appear in the

0097–6156/84/0247–0119$06.00/0

spectrum, and if $\Delta=3/4J$, methyl and methine carbon resonances appear upward and methylene carbon resonances appear downward. Since then, the INEPT method has proven to be quite useful when assigning complex ^{13}C NMR spectra. The purpose of this paper is to demonstrate the usefulness of the INEPT method for the characterization of low-density polyethylene and ethylene propylene copolymers.

Experimental

The low-density polyethylene(LDPE) was obained from the Rubber and Plastic Research Association(RAPRA). The polyethylene was dissolved in a mixture of 1,2,4-trichlorobenzene and benzene-d_6(4:1) at a concentration of 25%(w/v). The LDPE sample was the same as used in the previous work(16). The measurements were made at a temperature of 120°C.

The ethylene-propylene copolymer(EP) was provided by Japan Synthetic Rubber Co.(JSR). The propylene monomer content of this EP copolymer was reported to be 36.3 mole% by the manufacturer. The copolymer was dissolved in a mixture of o-dichlorobenzene and benzene-d_6(3:1) at a concentration of 15%(w/v). The copolymer measurements were made at a temperature of 100°C.

Carbon 13 NMR spectra were obtained using a JEOL GX-500 spectrometer with a quadruture detection operating at a frequency of 125.77 MHz in a pulse-Fourier transform mode. FID's were acquired with a 16 bit A/D converter and stored on 32 or 64 K memory locations with 32 bit word length. The chemical shift was measured from an internal standard, hexamethyldisiloxane, which was taken as 2.03 ppm from tetramethylsilane(TMS). Internal lock was provided by an addition of benzene-d_6.

Results and Discussion

Low-Density Polyethylene   Figure 1 contains an ^{1}H noise-decoupled ^{13}C spectrum of LDPE. A partial spectrum from 28.5 to 32.0 ppm is shown with reduced intensity just below the whole spectrum. This spectrum was obtained with a spectral range of 10 KHz and with 6837 FID accumulations at a temperature of 120°C. A pulse width of approximately 10 μs corresponding to a forty five degree pulse angle and delay times between pulses of 3.6 s were used during data accumulation. The FID's were stored on 32 K memory locations. As previously reported(4-16), there appear to be a number of peaks on both sides of the main methylene peak centered at 30.00 ppm. It is clearly demonstrated that this high magnetic field strength (corresponding to the proton resonance frequency of 500 MHz) remarkably enhances resolution and sensitivity. The major methylene resonances appearing in the range from 29.4 to 30.6 ppm consist of more than 10 peaks, and at least 80 peaks can be discerned in the whole spectrum.

Figure 2 shows (a) the methine and methylene region of the

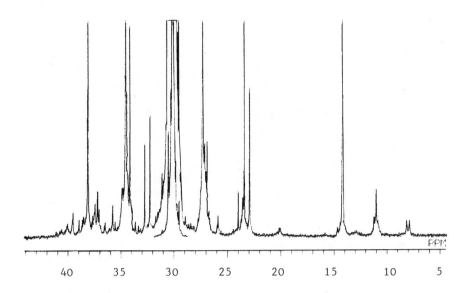

Figure 1. ¹³C NMR spectrum of the low-density polyethylene.

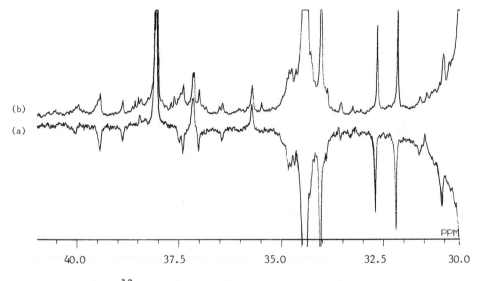

Figure 2. ¹³C INEPT spectrum with the delay time of Δ=3/4J(a) and the common spectrum(b) for the low-density polyethylene.

INEPT spectrum for LDPE, which was obtained with the delay time of $\Delta=3/4J$(6 ms) and 6000 FID accumulations and (b) the spectrum obtained by the common method, which is shown for comparison. In this figure one can easily distinguish methine resonances(upward peaks) from methylene resonances(downward peaks). Combined with the INEPT spectrum with the delay time $\Delta=1/2J$, all methyl, methylene, and methine resonances can be identified.

Of particular interest are resonances for the methine and methyl carbons, which are located at branch points and are branch end groups, respectively. The observed 15 different resonances for methine carbons suggest that at least 15 different branch types are present. Among these, the most intense resonance appears at 38.06 ppm and a second one occurs at 38.11 ppm which are probably associated with branch point methine carbons from butyl and longer chain branches. The highest field methine resonance appears at 31.48 ppm. Calculations using the Lindeman-Adams parameters(29) suggest that this peak may be associated with methine carbons attached to the methyl groups of any of 1,3-dimethyl branches, 1,3-paired methyl,ethyl or longer branches. In the vicinity of 33 ppm there appear three methine resonances, one of these is probably due to isolated methyl branches. The lowest field methine resonance appears at 40.74 ppm, below which we could not observe any methine resonances. Chemical shift and peak height intensity are listed Table 1. Peak height intensity is expressed in terms of percent of the total intensity. T_1 and nuclear Overhauser enhancement (NOE) measurements were made at this high field and at 120°C for the major methylene resonance, and we obtained T_1=2.1 s and NOE=2.6. Because no NOE measurements and no T_1 measurements for each resonance were made, intensity values listed in the table are only qualitative. In the range below 38.5 ppm, at least three methylene resonances can be found at 38.93, 39.48, and 40.08 ppm. These may originate from a C-α' methylene carbon in a 1,3-diethyl pair, a C-6 methylene carbon in a 5-ethylhexyl branch, a C-3 methylene carbon in a tetrafunctional methyl,propyl branch or any of those branch types illustrated by Axelson et al.(13).

In the high field methyl region, 14 methyl resonances can be identified as listed in Table I. The highest field methyl resonances appear at 7.85 ppm and 8.12 ppm and have been assigned to methyl carbons of tetrafunctional ethyl,butyl branches and tetrafunctional ethyl branches, respectively(16). An intense methyl resonance appearing at 14.1 ppm has been assigned to methyl terminals of butyl and longer chain branches when observed at lower magnetic fields. In the present study performed at a higher magnetic field, this resonance splits into two peaks at 14.10 and 14.15 ppm. Similar behavior was found for methine resonances at 38.06 and 38.11 ppm. This is a consequence of the enhanced resolution due to the high magnetic field strength, and indicates the possibility of the separation of methine and methyl resonances of butyl branches from those of long chain branches.

Table I. Chemical Shift Values and Peak Height Intensities of
the Low-Density Polyethylene

Methine Chemical shift	Intensity	Methyl Chemical shift	Intensity
31.48	0.10	7.85	0.06
33.12	0.02	8.12	0.06
33.27	0.02	10.74	0.04
33.58	0.06	10.89	0.06
35.76	0.13	10.90	0.06
37.17	0.17	11.00	0.19
37.20	0.17	11.16	0.07
37.87	0.05	11.22	0.08
38.06	0.91	12.88	0.02
38.11	0.58	14.10	0.57
38.46	0.07	14.15	1.12
38.53	0.08	14.39	0.05
39.89	0.03	14.66	0.03
40.02	0.05	20.03	0.03
40.74	0.02		

The resonance appearing near 20.0 ppm consists of two peaks, the one at 20.03 ppm is the methyl group resonance from methyl branches and the other at 20.14 ppm is associated with C-2 methylene of propyl branches. An integration of all methyl peaks indicates the presence of 24 methyl carbons per 1000 carbons, comparable to a value previously obtained of 19.8 (16).

Our results indicate that the previous assignments for methyl peaks are probably correct. In this study, it is possible to classify all resonances into methyl, methylene, and methine carbons. However, detailed assignments to specific structural units are still ambiguous, and more work is needed.

Ethylene Propylene Copolymer   The ¹H noise-decoupled ¹³C spectrum of an EP copolymer is shown in Figure 3. The spectrum was obtained over a spectral range of 8 KHz with 8352 FID accumulations at a temperature of 100°C. Forty five degree pulses and 6 s delay times between pulses were used. The FID's were stored on 64 K memory locations. As expected(17-27), more than 80 peaks appear in a range from 19 to 47 ppm.

Figure 4 shows the methine region of the INEPT spectrum observed with the delay time of $\Delta = 3/4J(a)$. Three thousand accumulations were stored on 32 K memory locations. The common spectrum(b) is also depicted in this figure. We can easily separate methine resonances(upward) from methylene

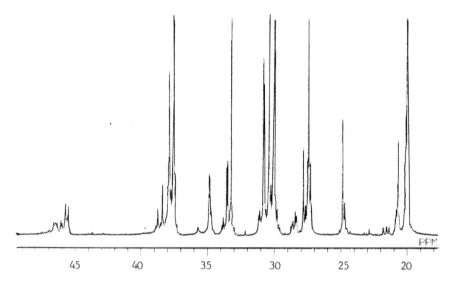

Figure 3. ^{13}C spectrum of the ethylene-propylene copolymer.

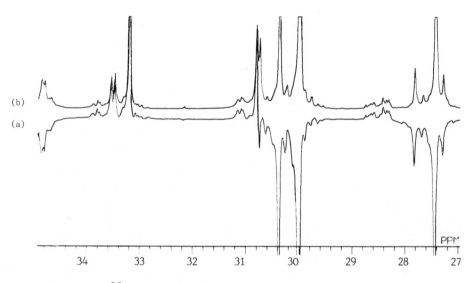

Figure 4. ^{13}C INEPT spectrum with the delay time of
$\Delta=3/4J$(a) and the common spectrum(b) for the ethylene-
propylene copolymer.

Table II. Chemical Shift Values and Peak Height Intensities of
the Ethylene-Propylene Copolymer

Methine Chemical shift	Intensity	Methyl Chemical shift	Intensity
28.33	0.7	19.97	13.2
28.37	0.7	20.00	8.6
28.44	0.9	20.07	4.0
28.54	0.3	20.18	0.9
28.61	0.5	20.43	0.2
28.67	0.4	20.59	0.7
28.76	0.3	20.66	3.2
30.81	6.5	20.70	1.7
31.07	0.8	20.74	1.0
31.09	0.9	20.81	0.9
31.16	0.7	20.85	0.7
32.87	0.1	20.90	0.5
32.98	0.3	21.36	0.3
33.05	0.3	21.54	0.3
33.10	0.3	21.57	0.3
33.20	15.8	21.78	0.2
33.32	0.9	21.82	0.3
33.48	2.7		
33.55	2.5		
33.77	0.5		
33.81	0.7		

resonances(downward).   Taking into account the INEPT spectrum
with the delay time of $\Delta=1/2J$, we identified all methine,
methylene, and methyl resonances.
     The methine resonances of this copolymer can be observed in 3
different regions,   28.3-28.7 ppm,   30.8-31.2 ppm,   and 32.8-
34.0 ppm.   Ray et al.(22) correctly assigned these methine
resonances. Of most interest is the differentiation of two peaks
appearing at 30.81 ppm and 30.75 ppm.   The INEPT spectra clearly
indicate that the one at 30.75 ppm can be assigned to methylene
carbons and the other at 30.81 ppm to methine carbons. According
to the assignments made by Ray et al.(22) the peak at lower field
is associated with the methine carbons of the PPE sequence and
the one at higher field to **YY** methylene carbons of the PEEP
sequence.   The methine resonances which appear at   28.3-28.4 ppm
have been assigned to the methine carbons in the PPP sequence.
     A spectral integration of the methyl region indicates that
the composition of propylene units is 37 mole percent   comparable
to the provided value.   Table II shows chemical shift values and

peak height intensities for methine and methyl resonances. Values
of T_1 and NOE for 8 major resonances were found to be   1.2-2.6 s
and 2.3-2.6, resepctively.

The present   study mostly confirms the results of previous
assignments   made   by Ray et al.(22) with respect to   the
identification of methyl, methylene, and methine peaks.   In
conclusion,   the   INEPT method provides a powerful technique for
characterization of polymers.

Acknowledgments   The authors wish to express their   appreciation
to Mr. K. Arai of JSR for providing the EP samples.

Literature Cited

1. Bovey,   F. A.   "High Resolution NMR of Macromolecules";
    Academic: New York, 1972.
2. Randall,   J.C.   "Polymer Sequence Determination:Carbon 13 NMR
    Method"; Academic: New York, 1978; Chap. 3, 6.
3. Randall, J. C. "Polymer Characterization by ESR and NMR";
    Woodward, A. E.; Bovey, F. A., Eds.; American Chemical
    Society: Washington, D.C., 1980; Symp. Ser. No.142, Chap. 6.
4. Dorman, D. E.; Otocka, E. P.; Bovey, F. A. Macromolecules
    1972, 5, 574-7.
5. Randall, J. C. J. Polym. Sci. Polym. Phys. Ed. 1973, 11, 275-
    87.
6. Randall, J. C. J. Polym. Sci. Polym. Phys. Ed. 1975, 13, 901-
    8.
7. Hama, T.; Suzuki, T.; Kosaka, K. Kobunshi Ronbunshu 1975, 32,
    91-6.
8. Bovey, F. A.; Shilling, F. C.; MaCrackin, F. L.; Wagner, H.
    L.   Macromolecules 1976, 9, 76-80.
9. Cudby, M. E. A.; Bunn, A. Polymer 1976, 17, 345-7.
10. Nishioka, A.; Ando, I.; Matsumoto, J. Bunseki Kagaku 1977,
    26,   308.
11. Cutler, D. J.; Hendra, P. J.; Cudby, M. A. E.; Willis, H. A.
    Polymer 1977, 18, 1005-8.
12. Randall, J. C. J. Appl. Polym. Sci. 1978, 22, 585-8.
13. Axelson, D. E.; Levy, G. C.; Mandelkern, L. Macromolecules
    1979, 12, 41-52.
14. Dechter, D. E.; G. C.; Mandelkern, L. J. Polym. Sci. Polym.
    Phys. Ed.   1980, 18, 1955-61.
15. Nishioka, A.; Mukai, Y.; Oouchi, M.; Imanari, M. Bunseki
    Kagaku 1980, 29, 774-80.
16. Ohuchi, M.; Imanari, M.; Mukai, Y,; Nishioka, A. Bunseki
    Kagaku 1981, 30, 332-8.
17. Carman, C. J.; Wilkes, C. E. Rubber Chem. Technol. 1971, 44,
    781.
18. Wilkes, C. E.; Carman, C. J.; Harrington, R. A. J. Polym.
    Sci.   Polym. Symp. 1973, 43, 237-50.
19. Tanaka, Y,; Hatada, K. J. Polym. Sci. Polym. Chem. Ed. 1973,
    11, 2057-68.

20. Carman, C. J.; Baranwal, K. C. Rubber Chem. Technol. 1975, 48, 705.
21. Carman, C. J.; Harrington, R. A.; Wilkes, C. E. Macromolecules 1977, 10, 536-44.
22. Ray, G. J.; Johnson, P. E.; Knox, J. R. Macromolecules 1977, 10, 773-8.
23. Randall, J. C. Macromolecules 1978, 11, 33-6.
24. Smith, W. F. J. Polym. Sci. Polym. Phys. Ed. 1980, 18, 1573-85.
25. Smith, W. V. J. Polym. Sci. Polym. Phys. Ed. 1980, 18, 1587.
26. Prabhu, P.; Schindler, A.; Theil, M. H.; Gilbert, R. D. J. Polym. Sci. Polym. Lett. Ed. 1980, 18, 389-94.
27. Randall, J. C.; Hsieh, E. T. Macromolecules 1982, 15, 1584-6.
28. Grant, D. M.; Paul, E. G. J. Am. Chem. Soc. 1964, 86, 2984-90.
29. Lindeman, L. P.; Adams, J. Q. Anal. Chem. 1971, 43, 1245-52.
30. Carman, C. J.; Tarpely, A. R.; Goldstein, J. H. Macromolecules 1973, 6, 719-24.
31. Morris, G. A.; Freeman, R. J. Am. Chem. Soc. 1979, 101, 760-2.
32. Doddrell, D. M.; Pegg, D. T. J. Am. Chem. Soc. 1980, 102, 6388-90.

RECEIVED October 31, 1983

NMR OF LIQUID POLYMERS

13C NMR in Polymer Quantitative Analyses

J. C. RANDALL and E. T. HSIEH

Phillips Petroleum Company, Research and Development, Bartlesville, OK 74004

Carbon 13 nuclear magnetic resonance can be used quantitatively in analyses of polymers to measure conveniently comonomer concentrations, average sequence lengths, run numbers and comonomer triad distributions. The identification of both short chain and long chain branches in polyethylene at concentrations of 1 per 10,000 carbon atoms has become feasible with the availability of improved probes and improved computer hardware/ software capabilities. Reviewed in this chapter are the methods and computations as well as the basic requirements for sound quantitative analyses: namely, correct choice of solvent, a consideration of concentration effect on line widths and satisfying nuclear Overhauser effects and spin lattice relaxation time requirements. Finally, the NMR generated structural information is put to use in correlations with polyethylene physical properties and measurements of number average molecular weight.

The importance of 13C NMR in determining polymer structure has become clearly evident through the abundance of literature on the subject during the past several years (1)(2). Sequence distributions involving as many as seven contiguous monomer units have been reported in studies of polymer configuration (3)(4) and triad information has been conveniently extracted from 13C NMR spectra of many copolymers (5-9). Much of the early emphasis in 13C NMR studies of polymers was on chemical shift assignments (2)(10)(11). After all, this was an essential prerequisite to any polymer 13C NMR structural analysis. Relaxation time and nuclear Overhauser effect measurements were also of considerable interest because of the information provided about polymer dynamics and requisite experimental considerations. The last area in NMR studies to be fully exploited was the quantitative method. It is natural for the polymer chemist to expect quantitative information once the power of the method in deciphering polymer structure has become evident.

0097–6156/84/0247–0131$06.25/0
© 1984 American Chemical Society

Improvements in instrumentation have led to increased spectral sensitivity and to more reliable methods for data integration. These improvements coupled with larger, more efficient and faster computers have given quantitative ^{13}C NMR applications the boost required for broad useage. A sensitivity of at least one structural unit per 10,000 carbon atoms is required for useful molecular weight and branching measurements and to allow species produced by oxidation and irradiation (12)(13) to be detected early in the process.

There are a number of considerations that must be addressed when formulating quantitative ^{13}C NMR procedures - these include solvent effects, spectral overlap, line widths, dynamic and nuclear Overhauser effects and detailed assignments. The steps required to develop sound quantitative methods will be the subject of this chapter. It is imperative that excellent quantitative methods be established so that NMR can be utilized in studies of polymer structure-property relationships. Polymer molecular structure needs to be related to the incipient solid state structure and ultimately to observed solid state physical properties such as density, flexural moduli, environmental stress cracking behavior, to name a few.

The usefulness of ^{13}C NMR in quantitative analyses of polymers is quite broad and covers a wide number of both addition and condensation polymers. For the sake of brevity, the present discussion concerning the development of quantitative procedures will be limited to ethylene-1-olefin copolymers and a few high density polyethylenes. These polyethylenes allow the typical NMR problems encountered to be explored while, at the same time, they severely test the instrumental dynamic range capabilities. Although polyethylenes have a relatively simple repeat unit structure, they may contain a variety of short chain branches, long chain branches and different types of end groups. In linear low density polyethylenes, the comonomer sequencing may offer overlap and assignment problems. The principles established in these quantitative studies can be usefully applied to other polymer systems of interest because the same basic considerations must be addressed. In the following sections, the choice of solvent, concentration effects, dynamic effects, ways to avoid assignment problems, sequence analyses, long chain branching and molecular weight measurements will be discussed in detail.

Experimental Variables in Quantitative NMR Studies of Polymers

Choice of Solvent. The most appropriate solvent for NMR studies of polymers would allow a range of polymer concentrations to be investigated, be free of overlap problems and hopefully provide a signal for internal lock. Not all of these conditions can usually be met as many high molecular weight polymers pose solubility problems and can be examined in only a limited number of solvents. Deuterium resonance is the typical choice for an internal lock signal on most modern NMR spectrometers. Unfortunately, the majority of available deuterated solvents are poor solvents for many addition polymers such as the polyolefins while it is generally possible to find a number of appropriate deuterated solvents for many of the condensation polymers. The chlorinated benzenes are the best solvents for NMR studies of polyethylenes and vinyl polymers. In particular, 1,2,4-trichlorobenzene has a high boiling point (213.5 C), poses no overlap problems in the 0-100 ppm (TMS) range, allows a significant range of polyethylene concentrations to be investigated and is stable over the long data acquisition periods required for ^{13}C NMR analyses. Usually a small amount of perdeuterobenzene can be added to 1,2,4-trichlorobenzene to provide a signal for lock purposes. If only a few hours of data accumulation are required, many superconducting magnet systems have sufficient stability to give excellent spectra without using any lock solvent. For overnight and longer runs, a lock solvent is preferred.

Effects of Polymer Concentration. A normal approach when attempting to detect low quantities of structural units in polyolefins is to prepare a solution as concentrated as possible for the NMR study. The polymer signal strength per free induction decay will improve as the concentration increases. A factor not often considered to be of importance in quantitative NMR studies of polymers is the effect of concentration upon resonance linewidth. For most polyolefins, the resonance linewidth increases as the concentration of the polymer is increased. Carbon 13 NMR spectra of a Marlex 6003 high density polyethylene (M_w = 140,000, M_n = 20,000) are shown in Figure 1a at a concentration of 50% by weight in 1,2,4-trichlorobenzene and in Figure 1b at a concentration of 15% by weight with the remainder of the experimental conditions being the same. (The nomenclature is given under Section 3.) The resonance linewidth at one-half height for the major recurring methylene resonance, $\delta^+\delta^+$, changed from approximately 1.0 Hz at 15 weight % to approximately 10 Hz at 50 weight %. Although the end groups resonances have larger areas at a concentration of 50% relative to 15%, the signal to noise is actually better for the spectrum with the more narrow linewidths. The spectrum in Figure 1a was obtained after 4,992 FID's while that in Figure 1b was obtained after 5,518 FID's. If the FID accumulation for the 15% by weight sample is allowed to continue to 20,683, the spectrum shown in Figure 2 is obtained. The signal to noise is now such that one structural unit in 10,000 carbon atoms can be detected. Evidence that indicates the presence of long chain branching has now been observed as indicated by the weak resonances for the methine, $\alpha\delta^+$ and $\beta\delta^+$ carbons for a long chain branch.

$$\begin{array}{cccc} \beta\delta^+ & \alpha\delta^+ & \alpha\delta^+ & \beta\delta^+ \end{array}$$
$$\sim CH_2\text{-}CH_2\text{-}CH\text{-}CH_2\text{-}CH_2\sim$$
$$|$$
$$CH_2$$
$$|$$
$$CH_2$$

Methine = 38.19 ppm (from TMS)

$\alpha\delta^+$ = 34.55

$\beta\delta^+$ = 27.20

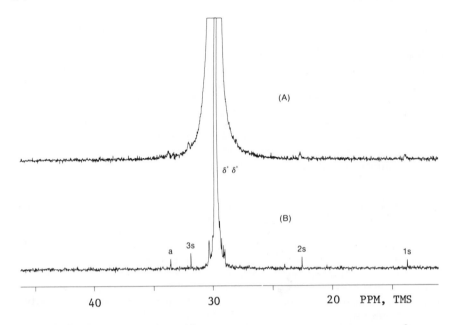

Figure 1   50.3 MHz ^{13}C NMR Spectra of 6003 PE at 125°C
             at (a) 50% by Weight and (b) 15% by Weight in
             1,2,4-Trichlorobenzene

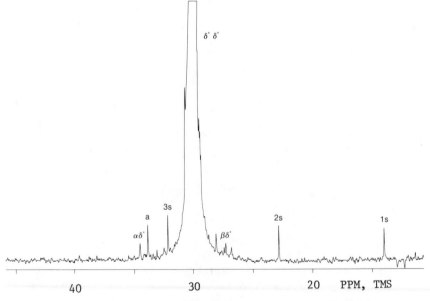

Figure 2   A 50.3 MHz ^{13}C NMR Spectrum of 6003 PE at
             125°C at 15% Weight in 1,2,4-Trichlorobenzene
             after an Accumulation of 20,683 FID's

From a comparison of peak heights, the long chain branching is approximately 1 per 10,000 carbons.

Another desirable result from spectra obtained with narrow line widths is that overlap is reduced not only for closely spaced resonances but also from the Lorentzian "tails" which influence small resonances within 10 ppm of a major resonance such as the $\delta^+\delta^+$ resonance in polyethylene. Integration of the end group resonances in Figure 2 leads to a number average molecular weight of 18,900. At polymer concentrations above 15% by weight, the measured number average molecular weight becomes a function of concentration as the overlap from the strong tail of the $\delta^+\delta^+$ resonance becomes more severe with increasing concentration. For polyethylenes, 15% by weight is an optimum concentration in terms of line width versus signal strength. For other polymers, the concentration effects would have to be established independently depending upon the proximity of the various resonances and the relative signal strengths. Accordingly, it may well be possible to work successfully with higher concentrations in polymer systems other than polyethylene.

Relaxation Time and Nuclear Overhauser Effects. It is well known that for 90° pulse angles, a pulse spacing of 5 x T_1 will ensure that 99% of rf excited nuclei will be fully relaxed between pulses (14). Such pulse spacings will ensure reliable quantitative results and are recommended although it is possible to obtain quantitatively reliable results with lower pulse angles. The following relaxation time (T_1) data was obtained for a 97/3 ethylene-1-hexene copolymer:

$$\underset{\sim CH_2-CH_2-CH_2-CH-CH_2-CH_2-CH_2-(CH_2)n-CH_2-CH_2-CH_2-CH_3}{\overset{\gamma\delta^+ \quad \beta\delta^+ \quad \alpha\delta^+ \quad 1.7 \quad 1.1 \quad 1.3 \quad 1.3 \quad 1.8 \quad 2.9 \quad 8 \quad 8 \quad 7}{}}$$

$$4B_4 \quad CH_2 \quad 1.1 \qquad \delta^+\delta^+ \quad 4s \quad 3s \quad 2s \quad 1s$$

$$3B_4 \quad CH_2 \quad 2.0$$

$$2B_4 \quad CH_2 \quad 4.4$$

$$1B_4 \quad CH_3 \quad 7$$

It is evident from these data that use of only the polymer backbone carbon resonances in a quantitative treatment could lead to an efficient NMR experiment. The spin-lattice relaxation times increase progressively for carbons towards the ends of the butyl branches and the chain ends. Similar spin-lattice relaxation time data was obtained from the inversion recovery (15) and progressive saturation (16) methods. The perfect 90° pulse angle experiment would have a pulse spacing of 40 seconds. This particular ethylene-1-hexene copolymer had a number average molecular weight of 5,300 and a weight average molecular weight of 31,100 which afforded an opportunity to obtain the end group spin-lattice relaxation times.

Nuclear Overhauser effects (14) can arise from energy transfer from the proton nuclear spin reservoir to the ^{13}C nuclear spin reservoir during proton heteronuclear spin decoupling. It is possible for the NOE to vary among various types of polymer carbons, particularly for protonated versus nonprotonated carbons in systems having highly restricted molecular mobilities.   Quite often full NOE's (~3.0) are observed for polymers because the dipole-dipole relaxation mechanism (14) will predominate because there is a restricted overall molecular mobility yet there are sufficient internal segmental motions to allow for a full NOE.   When NOE's do vary among polymer carbons of different types, there are several courses of action.   The NOE can be measured directly through gated decoupling experiments (17) and each observed resonance intensity can then be corrected for NOE differences.   If the signal to noise ratio is such that accurate NOE's cannot be obtained, it would be wise to utilize gated decoupling to preclude the possibility of differences in NOE contributions.

Nuclear Overhauser effect measurements were made for the 97/3 ethylene-1-hexene copolymer through gated decoupling experiments (17).   The NOE was full (~3.0) for all carbons including the ends of branches and end groups.   This result indicates that the dipole-dipole relaxation mechanism predominates suggesting, once again, a limited overall molecular mobility but indicating appreciable segmental motions.   This is a most fortunate occurrence since full NOE's greatly simplify the number of steps required in performing quantitative analyses on ^{13}C NMR data from polymers.

With NOE and T_1 information about the ethylene-1-hexene copolymer, we are in a position to design an efficient experiment for free induction decay (FID) accumulations.   Two experiments were performed: (a) a $90°$ pulse angle, (pulse width = 10.5 μs) a 15 second pulse delay and a 1.0 second acquisition time, and (b) ~$30°$ pulse angle, (pulse width = 3.5 μs) no pulse delay and a 1.0 second acquisition time. Carbon 13 NMR spectra are presented in Figure 3a for the $90°$ pulse angle experiment and in Figure 3b for the ~$30°$ pulse angle experiment. The observed peak heights and relative areas are presented in Table I. The number of FID's accumulated for the ~$30°$ pulse angle experiment was 14,400 while only 1,441 were accumulated for the $90°$ pulse angle experiment.   The ~$30°$ pulse angle experiment required four hours of instrument time whereas the $90°$ pulse angle experiment ran for 6.4 hours. Significant time savings can be realized with lower pulse angles when ^{13}C nuclei are present with substantially long relaxation times. As can readily be seen in Figure 3a, the signal to noise ratio is much better for the $90°$ pulse angle spectrum. (The pulse angle at 3.5 μs is likely less than $30°$. Approximately 1 μs is required to turn the pulse on.) The mole percent 1-hexene, using a method which will be presented shortly, was 3.1 for the $90°$ pulse angle experiment and 3.5 for the ~$30°$ pulse angle experiment. The methyl resonance with the longest T_1 was not used in this determination of mole percent 1-hexene although $2B_4$ was utilized. If one only uses resonances which have a T_1 less than 2.0 seconds, similar results are obtained with a value of 3.3 mole percent for the $90°$ experiment and 3.8 mole percent for the ~$30°$ experiment. These results indicate that a pulse spacing of less than 5 x T_1 will not lead to serious errors if the pulse spacing is still greater than 3 x T_1 for the slowest relaxing nucleus and most of the ^{13}C nuclei are totally

Table I

Relative Peak Heights and Integrated Areas from 50.3 MHz ^{13}C NMR Spectra of a 97/3 Ethylene-1-Hexene Copolymer Utilizing Different Pulse Angles and Pulse Spacings.

Line	PPM,TMS	Heights	Areas	Heights	Areas
1.	38.15	0.013	0.014	0.008	0.015
2.	34.55	0.019	} 0.048	0.017	} 0.054
3.	34.15	0.012		0.012	
4.	32.17	0.008	0.008	0.007	0.011
5.	30.47	0.021		0.019	
6.	29.98	0.831	} 0.860	0.848	} 0.844
7.	29.55	0.010		0.011	
8.	29.52	0.013		0.011	
9.	27.28	0.022	0.029	0.022	0.031
10.	23.36	0.019	0.014	0.017	0.016
11.	22.86	0.005	0.008	0.008	0.008
12.	14.08	0.019	} 0.020	0.012	} 0.022
13.	14.03	0.006		0.008	

satisfied dynamically by the experimental pulse spacing. Quantitatively, the results are similar for the two experiments utilizing different pulse angles.

The determination of mole percent 1-hexene in the previous example was based on relative areas measured by spectral integration. The use of peak heights gives mole percents 1-hexene of 2.8 from the 90° data and 2.3 from the ~30° data. Relative areas are normally recommended because any difference in resonance line widths will be taken into consideration through a use of relative areas.

Once a reliable ^{13}C NMR spectrum of a polymer is obtained, one must still face problems associated with resonance overlap and detailed assignments. If the polymer solution is too concentrated and strong signals are present, overlap from Lorentzian "tails" can lead to large errors in a quantitative determination. A satisfactory way to avoid some of these problems is given in the following section.

Method of Collective Assignments. One way to avoid errors introduced into a quantitative analysis of ^{13}C NMR data of polymers by spectral overlap and complex fine structure is to divide the observed spectrum into spectral regions sufficiently broad to contain overlapping resonances and fine structure. Closely spaced resonances will often result from a high chemical shift sensitivity to either configuration or comonomer sequences. Configurational sensitivities often lead to observable, but incompletely resolved pentads and heptads where the various pentads and heptads are grouped according to common central triads which are sufficiently resolved as separate multiplets. The methyl region of polypropylene is a good example of such behavior. As shown in Figure 4, sequence sensitivity is to both pentads and heptads yet the three basic triad centers are well resolved as separate complex multiplets. Under such circumstances a triad assignment is all that is required because of the necessary triad-pentad and pentad-heptad relationships. Two examples are given below:

$$\underline{mm} = m\underline{mm}m + m\underline{mm}r + r\underline{mm}r \qquad (1)$$

$$\underline{mmmm} = m\underline{mmmm}m + m\underline{mmmm}r + r\underline{mmmm}r \qquad (2)$$

The same principle applies to comonomer sequence chemical shifts. One needs only to identify the basic dyad in tetrads, hexads, etc., or the basic triad in pentads and heptads, etc. The final equations can be expressed in any adjacent complete distribution as desired through use of the appropriate necessary relationships (2) and as dictated by the number of independent spectral observations. The spectral region to be defined should be broad enough to include any unusual chemical shift behavior. An example best illustrates such an approach; the 97/3 ethylene-1-hexene copolymer shown in Figure 3 and an 83/17 ethylene-1-hexene copolymer shown in Figure 5 serve as useful prototypes. The spectral regions are defined below in terms of spectral range, sequence assignments and contributing carbons. The nomenclature is that used throughout where Greek symbols are used to denote the nearest branched carbons in both directions from a

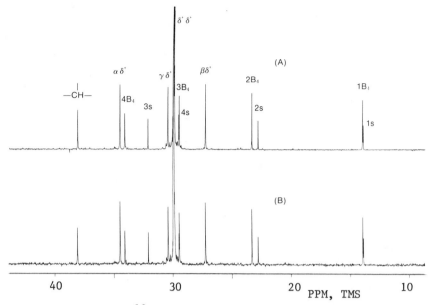

Figure 3   50.3 MHz ^{13}C NMR Spectra of a 97/3 Ethylene-1-Hexene Copolymer at 125° and 15% by Weight in 1,2,4-Trichlorobenzene after (a) a 90 Pulse Angle and 16 sec Pulse Interval and (b) a 30 Pulse Angle and 1 sec Pulse Interval

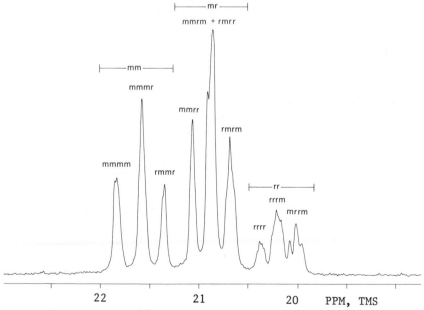

Figure 4   A 50.3 MHz ^{13}C NMR Spectrum of the Methyl Region of an Atactic Polypropylene at 125°C in 1,2,4-Trichlorobenzene (Sample courtesy of Walter Kaminsky, Univ. of Hamburg, W.Germany)

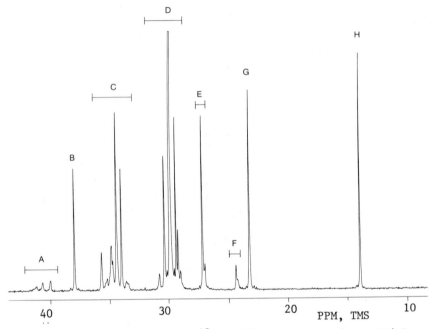

Figure 5   A 50.3 MHz ^{13}C NMR Spectrum of an 83/17
              Ethylene-1-Hexene Copolymer at 125°C and 15%
              by Weight in 1,2,4-Trichlorobenzene

methylene carbon of interest. The backbone methylene carbons in poly(1-hexene), for example, are all $\alpha\alpha$. (For consistency, the final equations are expressed in terms of triads only.)

<u>Region "A"</u>   39.5 to 42 ppm $\alpha\alpha$ CH_2 from HHHH, HHHE, EHHE

The $\alpha\alpha$ methylene carbons normally display a tetrad sensitivity. In this case, the H<u>HH</u>H $\alpha\alpha$ carbons are broadened by the presence of closely spaced resonances from sequences of different configurations (see Figure 5). Collectively, the entire region is simply proportional to the HH dyad concentration, that is,

$$\text{Total Area} \quad = T_A \quad = k\,[(\text{HHHH}) + (\text{HHHE}) + (\text{EHHE})] = k(\text{HH}) \qquad (3)$$

and, in terms of triads through the "necessary relationships" (2) becomes

$$T_A \quad = k\,[\,(\text{HHH}) + (1/2)\,(\text{EHH})\,] \qquad (4)$$

<u>Region "B"</u>   38.1 ppm   Methine from EHE sequences

$$T_B \quad = k\,(\text{EHE}) \qquad (5)$$

Region "C" is somewhat more complicated because it consists of both methine and methylene carbon resonances.

<u>Region "C"</u>   33 to 36 ppm   Methine carbons from EHH and HHH
$4B_4$ carbons from EHE, EHH and HHH
$\alpha\delta^+$ from E<u>H</u>EE and H<u>H</u>EE
$\alpha\gamma$ from E<u>H</u>EH and H<u>H</u>EH

and, finally,

$$T_C \quad = k\,[2\,(\text{HHH}) + 3\,(\text{EHH}) + 3\,(\text{EHE})] \qquad (6)$$

Region "D" contains the strong $\delta^+\delta^+$ resonance from the recurring methylene units and interferes with nearby resonances which originate from the $3B_4$ branch carbon, and the $\gamma\gamma$, $\gamma\delta^+$ backbone carbons close to a butyl branch.

<u>Region "D"</u>   28.5 to 31 ppm $\gamma\gamma$ from HEEH
$\gamma\delta^+$ from HEEE
$\delta^+\delta^+$ from $(\text{EEE})_n$
$3B_4$ from EHE, EHH and HHH

and

$$T_D = k\,[2\,(\text{EEE}) + (1/2)(\text{HEE}) + (\text{EHE}) + (\text{EHH}) + (\text{HHH})] \qquad (7)$$

where the following necesary relationships (7) were needed to obtain equation 7.

$$(EEE) = (1/2)\ \delta^+\delta^+\ +\ \ (1/4)\ \gamma\delta^+ \tag{8}$$

$$(1/2)\ (HEE) = (HEEH) + (1/2)(HEEE) \tag{9}$$

Regions "E" and "F" involve multiple resonances but from only one type of carbon in each case.

Region "E"   26.5 to 27.5 ppm $\beta\delta^+$ from E̲H̲E̲E̲ and H̲H̲E̲E̲

$$T_E = k(HEE) \tag{10}$$

Region "F"   24 to 25 ppm $\beta\beta$ from E̲H̲E̲H̲E̲, E̲H̲E̲H̲H̲ and H̲H̲E̲H̲H̲

$$T_F = k(HEH) \tag{11}$$

The HEH-centered pentads in region "F" are not completely resolved, thus the use of collective assignments is most valuable in this case where one actually needs only the HEH triad to complete the triad set.

Regions "H" and "G" contain single resonances from the first and second carbons of the butyl branch, respectively. Although they relate directly to the total 1-hexene content in the copolymer, they can be expressed as triads for solution of the complete set of simultaneous equations.

Region "G"   23.4 ppm $2B_4$ from "H"

$$T_G = k[(EHE) + (EHH) + (HHH)] \tag{12}$$

Region "H"   14.1 ppm $1B_4$ (methyl) from "H"

$$T_H = k[(EHE) + (EHH) + (HHH)] \tag{13}$$

Other problems can develop in the use of Equations 4 through 13 because overlap can occur from resonances from sources other than repeat unit resonances. For example, the allylic carbon resonance at 33.91 ppm which arises from a terminal vinyl group will probably overlap with the $4B_4$ branch methylene resonance for EHE at 34.13 ppm. In Figure 5, the saturated end group resonances are so weak that overlap with corresponding resonances from the butyl branches is not a problem. For lower molecular weight copolymers such as in Figure 3, corrections must be made where necessary in the equations describing the spectral region where the end group resonances reside. The use of equations for spectral regions A, B, D, F and G do not offer possible overlap problems with end group resonances and give the following relationships for the triad concentrations:

$$k\ (EHE) = T_B \tag{14}$$

$$k\ (EHH) = 2(T_G - T_B - T_A) \tag{15}$$

$$k\ (HHH) = 2T_A + T_B - T_G \tag{16}$$

$$k \, (HEH) = T_F \tag{17}$$

$$k \, (HEE) = 2(T_G - T_A - T_F) \tag{18}$$

$$k \, (EEE) = (1/2) \, (T_A + T_D + T_F - 2T_G) \tag{19}$$

The integrated areas from the ^{13}C NMR spectrum in Figures 3 and 5 lead to the following results for the triad distribution:

	83/17 Copolymer	97/3 Copolymer
(EHE)	0.098	0.031
(EHH)	0.053	0.000
(HHH)	0.022	0.000
(HEH)	0.043	0.000
(HEE)	0.164	0.061
(EEE)	0.620	0.908

The "E" spectral region was not used in the above analyses because phasing difficulties led to consistently high results. This will be a typical problem in ^{13}C NMR spectra where there is a strong dominant resonance close to substantially weaker resonances of interest.

A triad distribution is useful because it gives the relative concentrations of each of the possible connecting sequences. With this information, the comonomer concentrations, number average sequence lengths and the "run number" can be calculated as follows:

$$(H) = (HHH) + (EHH) + (EHE) \tag{20}$$

$$(E) = (EEE) + (HEE) + (HEH) \tag{21}$$

$$\text{Run Number} = (1/2)(HE) \tag{22}$$

$$= (EHE) + (1/2)(EHH) = (HEH) + (1/2)(HEE) \tag{23}$$

$$\text{Average "E" Sequence Length} = (E)/\text{Run Number} \tag{24}$$

$$\text{Average "H" Sequence Length} = (H)/\text{Run Number} \tag{25}$$

The concept of run number was first introduced by Harwood [18] and was defined as the number of like monomer runs per 100 polymer chain units. In the above equations, the run number is defined in terms of relative concentrations and can be converted to the Harwood definition by simply multiplying by 100.

The development of the concepts of run number, average sequence lengths and triad distributions would be of little more than academic interest if they could not be usefully applied. The concept of run number is most valuable in a consideration of the effect of comonomer content versus branch length in affecting polyethylene density. The following section utilizes the run number in a correlation with a number of polyethylene physical properties.

Structure-Property Relationships. The choice of comonomer and its density-concentration relationship has been of considerable interest among polymer chemists engaged in the preparation of ethylene-1-olefin copolymers. A graph of mole percent 1-olefin versus density for the 1-butene, 1-hexene and 1-octene comonomers in ethylene-1-olefin copolymers is presented in Figure 6. The run numbers were determined from ^{13}C NMR data using the appropriate method of collective assignments and integrated resonance areas. The densities were determined from a density gradient column. A clear trend can be seen in spite of the scatter in the data. As the mole percent 1-olefin increases, the density of the polymer decreases. An effect from branch length is also indicated. The 1-octene comonomer appears to be the most effective in reducing density while the 1-butene comonomer appears to be the least effective. The points for the ethylene-1-hexene copolymers are in between those for the 1-butene and 1-octene copolymers.

These same copolymers are presented in Figure 7, but this time the density is plotted versus the run number instead of mole percent 1-olefin. The run number gives the average number of comonomer disruptions per 100 chain units irrespective of the run length of the disrupting units. Thus, the tendency of a comonomer to "cluster" in contiguous sequences is taken into account. A dimer or trimer of contiguous comonomer sequences would count as a disrupting unit in the same way as an isolated comonomer unit. It is quite apparent in Figure 7 that the scatter of points associated with different branch lengths is much reduced. The density clearly decreases as the run number increases but the effect of branch length now is much less pronounced than in Figure 6. It would appear that the ethylene-1-olefin copolymer density is more dependent upon chain disruptions as indicated by the run number than upon the difference in branch lengths. The fact that 1-butene is less effective in reducing the density in an ethylene copolymer than either 1-hexene or 1-octene may be more related to its tendency to form "clusters" than to its shorter branch length. The 1-octene comonomer shows very little tendency to cluster because the relationship observed in the graph of mole percent versus polymer density is quite similar to that observed for run number versus density. The largest differences between the two graphs were shown by those ethylene-1-butene copolymers where the EBB and BBB sequences gave measureable concentrations.

The extent of clustering for any ethylene-1-olefin copolymer can be defined by dividing the run number by the mole percent 1-olefin. This result is called the "monomer dispersity" as shown below:

$$\text{Monomer Dispersity} = 100 \times [(1/2)(EH)/(H)] \qquad (26)$$

A value of 100 will be obtained for the monomer dispersity whenever the comonomer exists solely as an isolated unit. In Figures 6 and 7, the monomer dispersities of the 1-butene copolymers varied from 60 to 99, the 1-hexene copolymers varied from 85 to 99 and the 1-octene copolymers had monomer dispersities which varied from 96 to 100. This

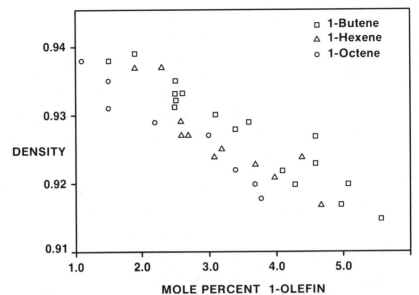

Figure 6   Mole Percent 1-Olefin Versus Density for a Series
of Ethylene-1-Butene, Ethylene-1-Hexene and
Ethylene-1-Octene Copolymers

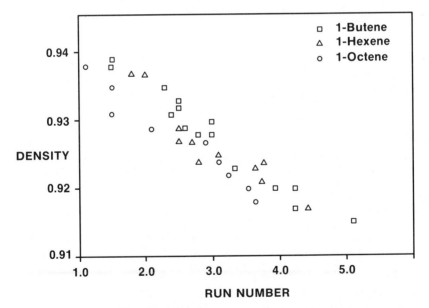

Figure 7   Run Number Versus Density for a Series of
Ethylene-1-Butene, Ethylene-1-Hexene and
Ethylene-1-Octene Copolymers

quite large difference in monomer dispersities suggests that an effect of branch length on reducing density could only be investigated through an examination of copolymers with similar (near 100) monomer dispersities. The points in Figures 6 and 7 which had monomer dispersities of 94 to 100 are replotted in Figure 8. Although the 1-octene points fall at the lower densities, it is very difficult to ascertain any significant difference between the ethylene-1-butene and ethylene-1-hexene copolymers. The scatter in the points, particularly at the higher densities, are related to differences in molecular weight and molecular weight distribution. This correlation will be repeated in a new study using samples from fractionated polymers.

For those copolymers where other physical property data was available, an attempt was made to correlate the run number with the flexural modulus and the tensile strength at yield. Results are given in Figures 9 and 10 for a limited number of points. This data should be obtained from carefully annealed samples or ones with similar thermal histories; however, a clear trend is seen in both figures. The run number does relate to crystallinity and to attendant physical properties such as flexural modulus and tensile strength.

With the detailed structural information provided through ^{13}C NMR analyses of polymers, the polymer scientist is in a position to determine physical property - structure relationships. Many excellent studies will be forthcoming during this decade, largely because of the development of NMR spectrometers with high sensitivity and sophisticated pulsing techniques which greatly simplify quantitative studies.

One final area for discussion is the use of NMR for the determination of polymer number average molecular weights. End group resonances were clearly visible in the ^{13}C NMR spectrum of the ethylene-1-hexene copolymer in Figure 3. An opportunity to determine polymer degrees of polymerization or number average molecular weights should not be overlooked.

Measurements of Number Average Molecular Weight. Carbon 13 NMR lends itself to a measurement of the degree of polymerization or number average molecular weight because a ratio of the number of repeat units to end units is all that is required for these determinations. The availability of floating point arithmetic during Fourier transformations and the use of double precision during data acquisition and manipulations have virtually eliminated problems associated with dynamic range. When evaluating the NMR technique for molecular weight measurements, it is prudent to use known molecular weight standards and NBS 1475 (M_w = 53,100; M_n = 18,300) is a good choice for such an investigation. NBS 1475 is reported to be a linear, high density (0.9784) polyethylene. A ^{13}C NMR spectrum is shown in Figure 11. A resonance at 33.91 ppm reveals the presence of vinyl terminal groups (10) although it is clear from the relative intensities that the saturated end groups are more abundant. The signal to noise ratio in Figure 11 is such that the end group and main chain resonances can be measured by spectral integration. This affords an opportunity to compare relative areas to relative peak heights as given below:

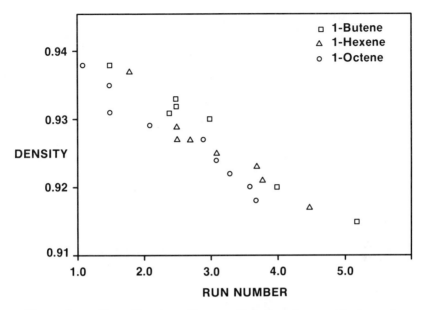

Figure 8   Run Number Versus Density for a Series of Ethylene 1-Butene, Ethylene-1-Hexene and Ethylene-1-Octene Copolymers Possessing Monomer Dispersities of 94-100

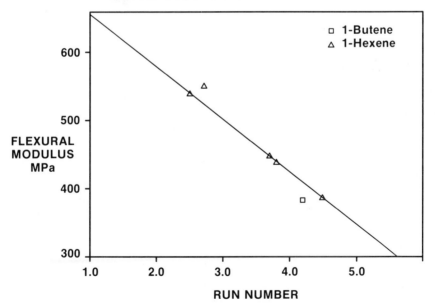

Figure 9   Run Number Versus Felxural Modulus for a Few Ethylene-1-Hexene and Ethylene-1-Butene Copolymers

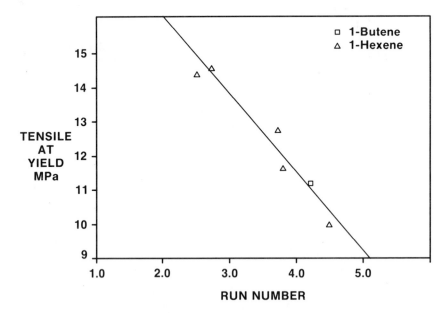

Figure 10   Run Number Versus Tensile Strength at Yield for a
               Few   Ethylene-1-Hexene   and   Ethylene-1-Butene
               Copolymers

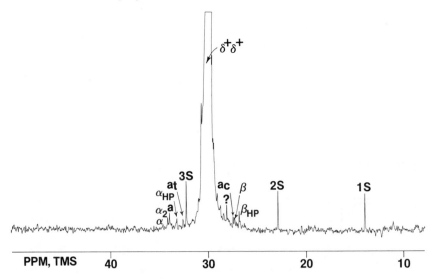

Figure 11   A 50.3 MHz ^{13}C NMR Spectrum of NBS 1475
               Polyethylene at 125°C and at 15 Weight Percent in
               1,2,4-Trichlorobenzene

NBS 1475 Polyethylene Standard

Carbon	Relative Areas*	Relative Peak Heights
1s	30.4	40.5
2s	29.1	50.7
3s	43.7	56.9
a[†]	16.0	19.4
$\delta^+\delta^+$	30,000.0	30,000.0

*The relative areas were determined by digital integration and normalized to $\delta^+\delta^+$ = 30,000 for comparison purposes.
[†]"a" is the allylic carbon resonance at 33.91 ppm for terminal vinyl groups.

It is apparent that the two NMR data sets (areas versus peak heights) will not lead to the same number average molecular weight. The following equation can be used to determine the number average molecular weight:

$$M_n = \frac{\text{Total Carbon Intensity}}{\text{Single Carbon Intensity}} \times 14 \qquad (27)$$

In the case of NBS 1475, the total carbon intensity and the single carbon intensity are given by

$$\text{Total Carbon Intensity} = (1s + 2s + 3s + a + \delta^+\delta^+) \qquad (28)$$

$$\text{Single Carbon Intensity} = (1/2)(\bar{s} + a) \qquad (29)$$

where $\bar{s}$ is the average intensity for 1s, 2s, and 3s. The number average molecular weight using Equations 27-29 and peak height data is 12,300, while a result of 16,700 is obtained using integrated areas. If one assumes that 3s is affected by the strong Lorentzian tail from the $\delta^+\delta^+$ resonance and is omitted from the calculation, a number average molecular weight of 18,400 is obtained from the area data. This latter result is in excellent agreement with the number average molecular weight reported by NBS.
     The difference in values obtained for the number average molecular weight according to the choice of peak intensity measurement arises from a difference in linewidths for the more mobile carbons. The following results were obtained for linewidths at one-half height for the observed NBS 1475 resonances at a solution concentration of 15% by weight in 1,2,4-trichlorobenzene and a temperature of 125°C.

Carbon	Linewidth at 1/2 Height
1s	1.1 Hz
2s	1.0
3s	0.9
a	(too low for a reliable measurement)
$\delta^+\delta^+$	1.2

The end group resonances are sufficiently more narrow (10-20%) to create a substantial error when determining repeat unit versus end group ratios by peak heights. Spectral integration is required and modern spectrometers equipped with digital integration can do the job.

Another reason that such good agreement was obtained between the NMR method and the gel permeation chromatography method used by NBS is that NBS 1475 has a fairly narrow molecular weight distribution (M_w/M_n = 2.9). It is possible that polyethylenes having very broad molecular weight distributions will give different number average molecular weights by the two methods. The NMR method is nondiscriminating as far as oligomers are concerned and will give the same end group resonances for a 36 carbon oligomer and a high molecular weight polymer. The GPC method is very insensitive to the presence of oligomers which could be present as a constituent in polyethylenes having broad molecular weight distributions. A comparison of number average molecular weights obtained from these two techniques therefore could be useful.

Conclusions

Until the last few years, our ability to synthesize and develop new polymers commercially far exceeded our ability to obtain detailed structural information, which was limited largely to comonomer concentrations, configuration, crystallinity and a determination of infrared sensitive functional groups. With the capability of Fourier transform ^{13}C NMR to detect connecting carbons between monomers of different types, we can now measure average sequence lengths, run numbers, dyad, triad and higher comonomer distributions, and branch length distributions. This information will give tremendous insight into structure-property relationships and solid state structures. With a sensitivity to structural entities at a level of one per ten thousand carbons or lower, ^{13}C NMR will be valuable in detecting the onset of degradation and linking reactions. A chapter on radiation induced structural changes in polyethylene appears later in this monograph. There can be little doubt that we are on the threshold of establishing structure-property relationships which were not considered possible a few short years ago.

Literature Cited

1. Bovey, F. A. "Chain Structure and Conformation of Macromolecules"; Academic Press: New York, 1982.

2. Randall, J. C. "Polymer Sequence Determination: Carbon-13 NMR Method"; Academic Press: New York, 1977.
3. Schilling, F. C.; Tonelli, A. E. Macromolecules 1980, 13, 270.
4. Randall, J. C. "Polymer Sequence Determination: Carbon-13 NMR Method"; Academic Press: New York, 1977; p 116.
    Sato, H.; Tanaka, Y. Chapter 12, this Monograph.
    Chen, T. K.; Harwood, H. J. Chapter 13, this Monograph.
5. Hsieh, E. T.; Randall, J. C. Macromolecules 1983, 15, 353.
6. Ray, G. J.; Spanswick, J.; Knox, J. R.; Serres, C. Macromolecules 1981, 14, 1323.
7. Ray, G. J.; Johnson, P. E.; Knox, J. R. Macromolecules 1977, 10, 773.
8. Hsieh, E. T.; Randall, J. C. Macromolecules 1982, 15, 1402.
9. Randall, J. C.; Hsieh, E. T. Macromolecules 1982, 15, 1584.
10. Carman, C. J.; Tarpley, A. R., Jr.; Goldstein, J. H. Macromolecules 1973, 6, 719.
11. Clague, A. D. H.; Van Broekhoven, J. A. M.; Blaauw, L. P. Macromolecules 1974, 7, 348.
12. Bovey, F. A.; Schilling, F. C.; Cheng, H. N. "Advances in Chemistry Series No. 169"; Allara, D. L.; Hawkins, W. L., Eds.; American Chemical Society: Washington, D.C., 1978; pp. 133-141.
13. Randall, J. C.; Zoepfl, F. J.; Silverman, J. Makromol. Chem., Rapid Commun. 1983, 4, 149.
    Randall, J. C.; Zoepfl, F. J.; Silverman, J. Chapter 16, this Monograph.
14. Farrar, T. C.; Becker, E. D. "Pulse and Fourier Transform NMR"; Academic Press: New York, 1969.
15. Gupta, R.; Ferretti, J.; Becker, E.; Weiss, G. J. Magn. Reson. 1980, 38, 447.
16. Freeman, R.; Hill, H. D. W. J. Chem. Phys. 1971, 54, 3367.
17. Freeman, R.; Hill, H. D. W.; Kaptein, R. J. Magn. Reson. 1972, 7, 327.
18. Harwood, H. J.; Ritchey, W. M. Polymer Letters 1964, 2, 601.
19. Randall, J. C. "Polymer Characterization by ESR and NMR", Woodward, A. E.; Bovey, F. A., Eds.; ACS SYMPOSIUM SERIES No. 142, American Chemical Society: Washington, D.C., 1980; pp. 93-118.

RECEIVED December 10, 1983

The Synthesis of Novel Regioregular Polyvinyl Fluorides and Their Characterization by High-Resolution NMR

RUDOLF E. CAIS and JANET M. KOMETANI

Bell Laboratories, Murray Hill, NJ 07974

We report the first synthesis of poly vinyl fluoride (PVF) having a pure head-to-tail (*isoregic*) sequence of monomer units. The procedure involves dechlorination of appropriate *isoregic* precursor polymers which can be chosen to either allow or prevent racemization of the product PVF. We have also prepared head-to-head, tail-to-tail (*syndioregic*) PVF by the copolymerization of ethylene with 1,2-difluoroethylene, and predominantly *syndiotactic* PVF by polymerization in an urea clathrate complex. The chemical microstructures of these polymers, as well as that of commercial regioirregular (*aregic*) PVF, have been analyzed by 188 MHz fluorine-19 NMR. Quantitative statistical descriptions of both the regiosequence and stereosequence distributions have been deduced from the NMR measurements. We show that free-radical homopolymerization of VF typically involves 11% of reverse monomer addition which results in *aregic* defect structures. The regiosequence distribution follows first-order Markov statistics, whereas the stereosequence distribution is Bernoullian.

The majority of vinyl polymers are regioregular with a head-to-tail sequence of monomer units (1), i.e. they are *isoregic* (2). Some of the most notable exceptions are the polymers obtained by free-radical addition reactions of the fluoroethylenes vinyl fluoride (VF), vinylidene fluoride (VF$_2$), and trifluoroethylene (F$_3$E), which incorporate significant amounts of head-to-head and tail-to-tail structural defects caused by reverse monomer addition (3).

We have devised a general procedure for the synthesis of *isoregic* polyfluoroethylenes, namely by the reductive dechlorination (or debromination) of appropriate precursor polymers in which head-to-head addition is sterically blocked by chlorine (or bromine) substituents. In this manner we have been able to prepare *isoregic* PVF, PVF$_2$, and PF$_3$E for the first time (4). The present report will deal with PVF, and details of the PVF$_2$ and PF$_3$E syntheses will be given in subsequent publications.

0097–6156/84/0247–0153$06.00/0
© 1984 American Chemical Society

PVF is a commercial polymer which exhibits unique weathering, mechanical, electrical, and chemical properties (5). It finds many applications in film form, including protective coverings on building and transportation materials, release coatings for molds, and glazing for solar panels. As we show here these materials typically contain about 11% of inverted monomer units, which create structural defects capable of influencing physical properties insofar as they may reduce crystallinity. One incentive for preparing pure *isoregic* polyfluoroethylenes is that they may exhibit improved properties compared to their *aregic* counterparts.

PVF has an additional source of structural irregularity owing to pseudoasymmetric $-CFH-$ methine carbons in the polymer backbone. In the commercial polymer these occur in nearly equal numbers of *meso* and *racemic* dyad pairs arranged in stereoirregular (*atactic*) sequences (6), as we shall see from the NMR analyses. An equilibrium stereosequence distribution is obtained after epimerization (7). When *isoregic* PVF is prepared by the reductive dechlorination of poly(1-chloro-1-fluoroethene), racemization takes place so that the stereosequence distribution may approach the equilibrium state. However racemization of the $-CFH-$ centers does not take place during the reductive dechlorination of poly(1-chloro-2-fluoroethene), and we contrast the tacticities resulting from these two alternative routes to *isoregic* PVF, and the synthesis of PVF in urea at low temperature which produces a more stereoregular polymer.

Heretofore the tacticity of PVF could not be measured accurately by fluorine-19 NMR owing to the overlap of resonances from *aregic* sequences with those from stereoconfigurational triads and pentads in *isoregic* sequences (8). The initial assignments proposed by Weigert (9) were made on the basis of rather questionable chemical shift analogies with the carbon-13 NMR spectrum of polypropylene. We are now in a position to deduce more definitive assignments from the spectra of the novel polymers described herein combined with accurate measurements of sequence probabilities.

Experimental

Commercial PVF was obtained from Aldrich Chemical Company. PCR Research Chemicals Inc. supplied 1-chloro-1-fluoroethene (vinylidene chlorofluoride, VCF), which was degassed and distilled under high vacuum and sealed in glass ampules with 1 mol% of azo*bis*(isobutyronitrile) initiator. Polymerization took place at 45°C for 24 h with 95% conversion to the polymer PVCF. Complete details on VCF polymerizations over a range of temperatures will be published separately. The monomer 1-chloro-2-fluoroethene (CFE) was likewise obtained from PCR and vacuum distilled. It was polymerized for 5 days at 60°C in a sealed ampule with 1 mol% of acetyl peroxide to yield 89% of PCFE.

The Matheson Company Inc. provided VF monomer, which was passed through a column packed with silica gel and molecular sieves (for removal of inhibitor and water), and distilled under vacuum. Urea (Fischer, certified ACS) was recrystallized three times from anhydrous ethanol, and 250 mmol was placed in a bulb attached to a vacuum manifold. Then 160 mmol of purified VF was added to the bulb, followed by a trace of methanol (0.2 ml) to initiate complex formation (10). The bulb was held at -65°C for 11 days, during which time the urea crystals became opaque as the VF-urea clathrate complex formed.

The clathrate complex was exposed to cobalt-60 γ radiation for 24 h at -80°C for an absorbed dose of 6.5 Mrad to yield 16% of PVF after the urea was washed away with water. PCFE was prepared in urea in a similar manner.

Ethylene (Matheson, CP grade) and 1,2-difluoroethene (mixed *cis* and *trans* from PCR) were sealed as an equimolar mixture in a heavy-wall ampule and exposed to cobalt-60 γ radiation (effective dose rate = 0.27 Mrad/h) for 20 h at -80°C. A poor yield (ca. 1%) of a waxy solid was obtained.

Essentially complete reductive dechlorination of the precursor poly(chlorofluoroethylene)s to PVF was achieved after 24 h at 60°C in tetrahydrofuran with a molar excess of tri(n-butyl)tin hydride and 1 mol% of azo*bis*(isobutyronitrile) (11). The reaction mixture was homogeneous until dechlorination was almost complete, after which the mixture turned cloudy owing to the insolubility of PVF.

Usually less than 0.1% of chlorine remained after the above treatment according to analysis by X-ray fluorescence spectroscopy. Fluorine was not removed from the polymer owing to its inertness compared to chlorine (12). An elemental analysis performed by Galbraith Laboratories Inc. confirmed these findings (anal. calcd for C_2H_3F: C, 52.2; H, 6.5; F, 41.3; found: C, 52.2, H, 6.2; F, 41.6). More vigorous reducing conditions (24 h at 90°C in dimethylformamide solution) caused yellowing of the polymer, probably due to dehydrohalogenation.

Fluorine-19 NMR data were acquired at a frequency of 188.22 MHz with a Varian XL-200 spectrometer. Typically, 100 transients were accumulated from a 5% polymer solution by volume in dimethylformamide-d_7 placed in a 5 mm sample tube at 120°C with internal hexafluorobenzene as a reference (Φ = 163 ppm). A sweep width of 8000 Hz was used with 8 K computer locations (acquisition time 0.5s) and a 5.0 s delay between 90° pulses (9.0 μs duration). Proton heteronuclear coupling was removed by broad-band irradiation centered at 200 MHz. A modified Bruker WH-90 spectrometer allowed carbon-13 NMR spectra to be obtained with simultaneous proton and fluorine-19 broadband decoupling (13).

Results and Discussion

Microstructures of Poly(chlorofluoroethylene)s   The carbon-13 NMR spectrum of PVCF consists of a $-CH_2-$ resonance at 54.1 ppm and a $-CFCl-$ resonance at 108.8 ppm. There is no splitting of these lines due to tacticity, nor are there any other resonances to indicate the presence of regioirregular monomer sequences. However the polymer is stereoirregular, as shown by the fluorine-19 NMR spectrum in Figure 1. There are three principal resonances spread by 3 ppm owing to triad stereosequences, with some pentad fine structure which is barely resolved.

The triad components have been assigned to syndiotactic (rr), heterotactic (mr), and isotactic (mm) sequences in order of increasing field. Peak areas are consistent with a Bernoullian sequence distribution having a p(m) value of 0.49 (6).

The fluorine-19 NMR spectrum of PCFE appears far more complicated. Figure 2 shows the spectra from two PCFE samples, one prepared at 60°C (a) and one prepared at -80°C in urea (b). Each backbone carbon is a pseudoasymmetric center in PCFE, compared to every second carbon in PVCF, so that the dispersion of fluorine-19 chemical shift from stereoirregularity is much larger. This dispersion is almost 15 ppm for PCFE, and is similar to the spread observed in the fluorine-19 NMR spectrum of poly(1,2-difluoroethene) (14).

Polymerization of CFE in urea at -80°C has a mild stereoregulating influence compared to bulk polymerization at 60°C, although the spectra in Figure 2 show that both polymers are substantially atactic. Three major triad resonances can be discerned more readily in spectrum (b), and these are assigned to mm, mr, and rr stereosequences as shown simply by analogy with the poly(1,2-difluoroethene) assignments (14,15). At present we have no basis to make definitive assignments, but those indicated are at least consistent with the known syndiotactic bias exerted by urea on the polymerization of complexed vinyl monomers (10).

General Features of PVF Spectra The proton- and fluorine-decoupled 22.62 MHz carbon-13 NMR spectra of PVF prepared from PVCF (a) and commercial PVF (b) are shown in Figure 3. There are five additional peaks present in spectrum (b) from the commercial polymer which are absent in spectrum (a). These are due to aregic monomer sequences, which have been assigned according to Tonelli et al. (16). Monomer sequence triads are resolved and are denoted by the binary regiosequence pentad notation in Table 1 (1 = CFH, O = CH_2).

The absence of these aregic defect resonances in spectrum (a) confirms that PVF obtained by the reductive dechlorination of PVCF is indeed isoregic. The 10101 and 01010 peaks have fine structure due to stereoirregularity, although this is not well resolved by carbon-13 NMR (13). Fluorine-19 NMR is more sensitive to the microstructure of PVF, as indicated in Figure 4, which compares the spectra from the above PVF samples at 84.66 MHz. Most of the resonances from aregic sequences occur at high field (spectrum b, 190 to 200 ppm), with the major peak intensity concentrated in the isoregic resonances which are split into a stereosequence triad from 179 to 183 ppm. Again it is evident from spectrum (a) that PVF derived from PVCF is regioregular, although stereoirregular. We now examine the tacticity of PVF more closely.

Stereosequence Microstructure of PVF Figure 5 shows in detail the 188 MHz fluorine-19 NMR signals from isoregic sequences in commercial PVF (a) and PVF prepared at -80°C in urea (b). The resolution at the higher spectrometer frequency is better (cf Figure 4), and it is further improved at 282 MHz (17). The resonance lines are much broader from the urea polymer because this material does not dissolve completely in dimethylformamide. We believe that crosslinking takes place through urea molecules during gamma irradiation, since PVF prepared under similar conditions but without urea is soluble.

The major splitting of the isoregic resonance into the three peaks is due to triad stereosequences, as noted above. Assuming that polymerization at low temperature in urea favors syndiotactic propagation, we assign these peaks to isotactic, heterotactic, and syndiotactic sequences in order of increasing field as

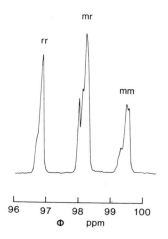

Figure 1.
188 MHz Fluorine-19 NMR of PVCF, $(CH_2-CClF)_n$, observed at 20°C as a 10% solution in acetone-d_6.

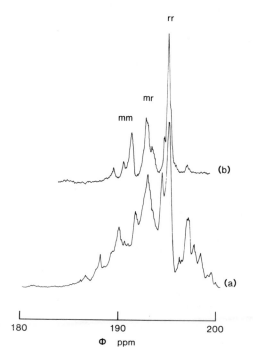

Figure 2.
188 MHz Fluorine-19 NMR spectra of PCFE, $(CHCl-CHF)_n$, prepared at (a) 60°C and (b) -80°C in urea, and observed at 60°C in dioxane-d_8.

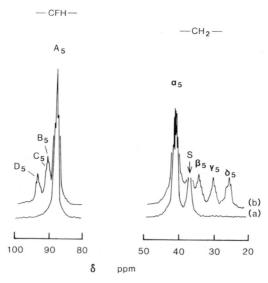

22.62 MHz Carbon-13 NMR spectra obtained with broad-band proton and fluorine decoupling of PVF obtained by reductive dechlorination of (a) PVCF and (b) commercial PVF. Spectra observed at 100°C with dimethylsulfoxide-d_6 solvent (S = solvent peak). The peak notation is explained in Table I.

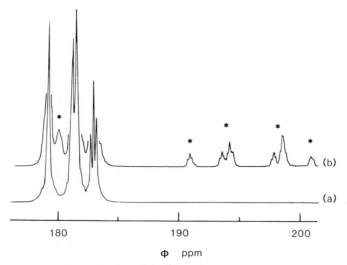

Figure 4.   84.66 MHz Fluorine-19 NMR spectra of PVF obtained by reductive dechlorination of (a) PVCF and (b) commercial PVF. The peaks marked with an asterisk are due to aregic sequences not present in the pure isoregic sample (a).

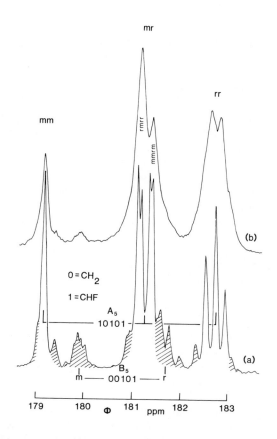

<u>Figure 5.</u>   Detailed expansion of the 188 MHz fluorine-19 resonances from the A_5 (10101) and B_5 (00101) sequences in (a) commercial PVF and (b) PVF prepared at -80°C in urea.

shown. These assignments are supported by calculations published by Tonelli et al. (15). Peak areas show that the probability of meso dyad formation, $p(m)$, is 0.46 for the commercial PVF and 0.35 for the urea PVF. The triad stereosequence distribution is Bernoullian for both polymers.

Fine structure from pentad stereosequences is clearly resolved in spectrum (a), although there is some overlap from peaks due to the 00101 (B_5) regiosequence. The stereosequence fine structure is seen more clearly in spectra from the pure isoregic PVF samples, which are shown in Figure 6. The spectra of PVF derived from PCFE (a) and PVCF (b) are virtually identical, and show that the stereosequence distributions are Bernoullian with $p(m)$ values of 0.49 and 0.48, respectively. The pentad assignments have been made according to these statistics by examining relative peak intensities and by comparison with Figure 5(b), where the pronounced syndiotactic bias distinguishes ambiguous cases (e.g. rmrr and mmrm) which arise when $p(m)$ is very close to 0.5. The weak peaks denoted by arrows in Figure 6(a) are probably due to end groups, since the precursor polymer PCFE had a relatively low degree of polymerization.

The bar graph in Figure 7 compares the triad stereosequence probabilities for the PVF samples examined here with values calculated for an equilibrium stereochemical distribution which would result from epimerization at 50°C (18). All PVF samples, except for the urea polymer, rather fortuitously have nearly an equilibrium stereosequence distribution. Furthermore, both the precursor PVCF and final PVF derived therefrom have nearly identical $p(m)$ values and stereosequence distributions, so that the effect of racemization during reductive dechlorination (19) is not apparent.

Regiosequence Microstructure of PVF In this final section we will be concerned with the regiosequence distributions in the aregic PVF samples. The A_5 and B_5 regiosequence pentads (Table 1) were identified in Figure 5(a), and we now examine the high-field resonances corresponding to C_5 and D_5 sequences. These are shown in Figure 8 for commercial PVF (a), urea PVF (b), and the copolymer of ethylene with 1,2-difluoroethene which serves as a model for syndioregic PVF (c).

There are six principal resonances from the two regiosequences C_5 and D_5 owing to further splitting from stereochemical effects. This splitting is larger when the $-CHF-$ carbons are adjacent (11) than when they are two bonds apart (101). We saw that m dyads are less likely than r dyads in the 101 sequence of urea PVF, so that the peaks denoted by the arrows in Figure 8(b) can be associated with m dyads. The two major resonances from the syndioregic polymer (c) establish the 00110 sequence assignments, with due regard for the complication of these signals by end group effects in the copolymer which has a very low degree of polymerization. Finally, we have proven that the B_5 and C_5 regiosequences must have equal probability, irrespective of the sequence statistics, whereas peak D_5 will have a different probability except in the special case of Bernoullian statistics (2).

Given the above considerations we arrive at the consistent set of assignments shown in Figure 8(a), which are supported by the calculations of Tonelli et al.

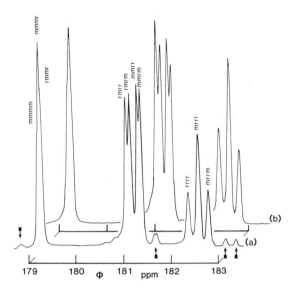

Figure 6.   Detailed expansion of the 188 MHz fluorine-19 resonances from isoregic PVF samples prepared by reductive dechlorination of (a) PCFE and (b) PVCF. The minor peaks marked with arrows in spectrum (a) are most likely due to end groups.

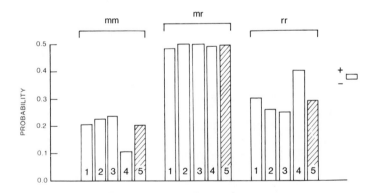

Figure 7.   The isotactic (mm), heterotactic (mr), and syndiotactic (rr) triad stereosequence probabilities for (1) commercial PVF, (2) PVF derived from PVCF, (3) PVF derived from PCFE, (4) urea PVF and (5) PVF epimerized at 50°C (18). Note that the observed heterotactic probability is mr + rm.

Table I. Set of Nonequivalent 1- and
O-Centered Regiosequence Pentads
(O = CH$_2$, 1 = CHF)

A$_5$	10101	α_5	01010
B$_5$	00101	β_5	11010
C$_5$	10110	γ_5	01001
D$_5$	00110	δ_5	11001

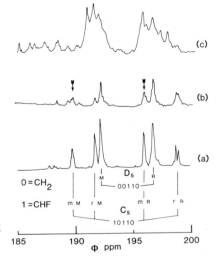

Figure 8.
Detailed expansion of the 188 MHz
fluorine-19 resonances from the
C$_5$ (10110) and D$_5$ (00110) sequences
in (a) commercial PVF, (b)
urea PVF, and (c) the copolymer model
for syndioregic PVF.

(15). These differ in detail from the assignments reported by Weigert ($\underline{9}$) and Görlitz et al. ($\underline{8}$). The present assignments are summarized in Table II.

The probabilities of the regiosequence pentads for commercial PVF and urea PVF are shown in Table III. For the former sample it is apparent simply by inspection that the regiosequence distribution is not Bernoullian, since $P_{obs}(C_5)$ and $P_{obs}(D_5)$ are different ($\underline{2}$). The distributions conform to first-order Markov statistics, characterized by two reactivity ratios r_0 and r_1, where $r_0 = k_{00}/k_{01}$ and $r_1 = k_{11}/k_{10}$ (k_{ij} is the rate constant for monomer addition to terminal radical i which generates the new terminal radical j). The present pentad data is insufficient to check the validity of this model, but it is unlikely that there is any deviation, as the same model has been tested and found adequate to describe the regiosequence distribution in PVF$_2$ ($\underline{2}$).

Table IV shows the reactivity ratios r_0 and r_1 derived from the probabilities in Table III in accord with a first-order Markov model ($\underline{2}$), where it is assumed that the more likely propagating terminal radical structure is 1 ($-$CHF·) and not 0 ($-$CH$_2$·). This assumption is consistent with gas phase reactions of VF with mono-, di-, and trifluoromethyl radicals, which add more frequently to the CH$_2$ carbon than to the CHF carbon ($\underline{20}$). The reactivity ratio product is unity if Bernoullian statistics apply, and we see this is not the case for either PVF sample, although the urea PVF is more nearly Bernoullian in its regiosequence distribution. Polymerization of VF in urea at low temperature also reduces the frequency of head-to-head and tail-to-tail addition, which can be derived from the reactivity ratios according to %defect $= 100(1 + r_0)/(2 + r_0 + r_1)$. Our analysis of the fluorine-19 NMR spectrum shows that commercial PVF has 10.7% of these defects, which compares very well with the value of 10.6% obtained from carbon-13 NMR ($\underline{13}$). Therefore the values of 26 to 32% reported by Wilson and Santee ($\underline{21}$) are in error.

Conclusions The free-radical polymerization of VF produces a regioirregular and stereoirregular polymer. The degree of irregularity can be reduced by low-temperature polymerization in urea. Isoregic PVF can be prepared by reductive dechlorination of either PVCF or PCFE. Preliminary studies by DSC show that the isoregic PVF melts at 217°C compared to 195°C for aregic PVF. One might expect that a stereoregular isoregic PVF would be even more crystalline.

Whereas the stereosequence distribution in isoregic and aregic PVF is nearly ideal random (Bernoullian with p(m) = 0.5), the latter has a first-order Markov regiosequence distribution. Accordingly the monomer sequence isomerism in PVF cannot be described by a single parameter such as the % defect, and requires two reactivity ratios for complete specification.

Table II.	Assignment of the 188 MHz Fluorine-19 NMR Spectrum of Commercial PVF to Regio-and Stereo-Sequence Pentads (observed in dimethylformamide at 120°C, internal hexafluorobenzene = 163 ppm, 11 stereochemical dyads = M and R, 101 stereochemical dyads = m and r).

Regiosequence	Φ ppm
A_5 (10101)	179.1 (mmmm), 179.1 (mmmr), 179.1 (rmmr), 181.0 (rmrr), 181.1 (rmrm), 181.3 (mmrr), 181.4 (mmrm), 182.5 (rrrr), 182.7 (mrrr), 182.9 (mrrm).
B_5 (00101)	179.8 (m), 181.7 (r)
C_5 (10110)	189.8 (mM), 191.8 (rM), 196.0 (mR), 198.8 (rR).
D_5 (00110)	192.3 (M), 196.8 (R).

Table III.	Observed Probabilities of Regiosequence Pentads ($P_{obs}(S_5)$) as defined in reference $\underline{2}$) for Aregic PVF Samples

Sequence	$P_{obs}(S_5)$	
	Commercial PVF	Urea PVF
A_5	0.349	0.398
B_5	0.047	0.034
C_5	0.047	0.034
D_5	0.057	0.034

Table IV.	Reactivity Ratios and Percent Defect (head-to-head and tail-to-tail) for Aregic PVF Samples

Parameter	Commercial PVF	Urea PVF
r_0	0.033	0.068
r_1	7.62	12.7
$r_0 \cdot r_1$	0.25	0.86
% defect	10.7	7.2

Literature Cited

1. Bovey, F. A. "Chain Structure and Conformation of Macromolecules"; Academic: New York, 1982; chap. 6.

2. Cais, R.E.; Sloane, N. J. A. Polymer, 1983, 24, 179-87.

3. Koenig, J. L. "Chemical Microstructure of Polymer Chains"; Wiley-Interscience: New York, 1980, chap. 9.

4. Cais, R. E.; Kometani, J. M. Proc. Org. Coat. Appl. Polym. Sci. Div. Amer. Chem. Soc., 1983, 48, 216-20.

5. Brasure, D. E. "Poly(Vinyl Fluoride)" in Kirk-Othmer Encyclopedia of Chemical Technology; Wiley-Interscience: New York, 1980, 3rd. Edition, Grayson, M., Exec. Ed., 11, 57-64.

6. Bovey, F. A. "High Resolution NMR of Macromolecules"; Academic: New York, 1972; chap. 3.

7. Suter, U. W. Macromolecules, 1981, 14, 523-8.

8. Görlitz, M.; Minke, R.; Trautvetter, W.; Weisgerber, G. Angew. Makromol. Chem. 1973, 29/30, 137-62.

9. Weigert, F. J. Org. Magn. Reson, 1971, 3, 373-7.

10. White, D. M. J. Amer. Chem. Soc., 1960, 82, 5678-85.

11. Hjertberg, T.; Wendel, A. Polymer, 1982, 23, 1641-5, and references therein.

12. Carlsson, D. J.; Ingold, K. U. J. Amer. Chem. Soc., 1968, 90 7047-55.

13. Schilling, F. C. J. Magn. Reson., 1982, 47, 61-67.

14. Cais, R. E. Macromolecules, 1980, 13, 806-8.

15. Tonelli, A. E.; Schilling, F. C.; Cais, R. E. Macromolecules, 1982, 15, 849-53.

16. Tonelli, A. E.; Schilling, F. C., Cais, R. E. Macromolecules, 1981, 14, 560-4.

17. Lin, Fu-Mei Chen, Ph.D Thesis, The University of Akron, Ohio, 1981.

18. Tonelli, A. E., personal communication.

19. Davies, D. I.; Parrott, M. J. "Free Radicals in Organic Synthesis"; Springer-Verlag: New York, 1978; p. 18.

20. Sloan, J. P.; Tedder, J. M.; Walton, J. C. J. Chem. Soc. Perkin II, 1975, 1846-50.

21. Wilson, C. W.; Santee, E. R. J. Polym. Sci. C, 1965, 8, 97-112.

RECEIVED September 22, 1983

The Composition and Sequence Distribution of Dichlorocarbene-Modified Polybutadiene by ^{13}C NMR

CHARLES J. CARMAN, RICHARD A. KOMOROSKI, and SAMUEL E. HORNE, JR.[1]

The B.F. Goodrich Research and Development Center, Brecksville, OH 44141

Carbon-13 NMR is being used to characterize the microstructure of a variety of chlorine-containing polymers. Among these are homopolymers, copolymers, and chemically modified polymers. In the last category is the series of polymers obtained by addition of dichlorocarbene to the double bonds of polybutadiene. Here we use ^{13}C NMR to examine a number of dichlorocarbene adducts of <u>cis</u>- and <u>trans</u>-polybutadiene prepared in a two phase system with phase transfer catalysis. Monomer compositions, comonomer sequence lengths, and stereochemical information are obtained for the resulting polymers. The polymers examined here were stereochemically pure and were treated as simple copolymers. Samples prepared using aqueous NaOH, CHCl$_3$ and phase transfer catalyst can be described as essentially random copolymers over the entire range of monomer composition. Samples prepared using solid instead of aqueous alkali metal hydroxides contain a higher fraction of blocked units than a polymer of comparable composition prepared using aqueous NaOH. This blockiness can coincide with the presence of two glass transition temperatures and a two-phase morphology. Fractionation of a substantially blocked sample yielded a chlorine-poor fraction which was a random copolymer and a chlorine-rich fraction which was more blocked than the original unfractionated material.

Portions of this work are reprinted from: R. A. Komoroski, S. E. Horne, Jr., and C. J. Carmen, *J. Polym. Sci., Polym. Chem. Ed., 21,* 89 (1983), copyright 1983, John Wiley and Sons, Inc. Reprinted by permission of John Wiley and Sons, Inc.

[1] Current address: Polysar, Inc., Stow, OH 44224.

^{13}C NMR has long been the technique of choice for the characterization of the molecular structure of homopolymers and copolymers (1). Recently, it has been used with success to study chlorine-containing polymers and copolymers (1-6). Among these are poly(vinyl chloride) (PVC) and various copolymers of vinyl chloride and other monomers both with and without chlorine. It has also proven to be powerful for characterizing modified polymers, in particular the products obtained from chlorination of poly-(vinyl chloride) (7,8) and polyethylene (9). Although both chlorinated PVC and chlorinated polyethylene are simple in their basic composition, consisting predominantly of CCl_2, $CHCl$, and CH_2 groups, the many possible arrangements of these groups produce very complex macromolecules. Both sequence distribution and CHCl configuration must be considered.

A more simple modified polymer is characterized in this report. Dichlorocarbene, $:CCl_2$, generated in situ, can add to polydienes to produce novel polymer containing dichlorocyclopropane rings at the points of addition (10-13). The reaction is illustrated below for cis-polybutadiene.

Polymers with a relatively broad range of physical properties can be prepared by varying the extent of reaction and hence chlorine content. These modified polymers can be viewed as a series of copolymers of varying comonomer composition and treated as such with regard to sequence distribution.

We have examined the microstructure of a number of dichloro-carbene adducts of both cis- and trans-polybutadiene using ^{13}C NMR spectroscopy. Samples were prepared in a two phase system where dichlorocarbene was generated by the reaction of either concentrated aqueous or solid alkali metal hydroxide with chloroform in the presence of a phase transfer catalyst (14). Monomer compositions and sequence lengths were obtained as for true copolymers and were correlated with glass transition temperature and phase morphology.

Results

In Table I are some pertinent data for a number of dichlorocarbene adducts of cis- and trans-polybutadiene. Figure 1 shows the ^{13}C NMR spectra of three cis-polybutadiene-based adducts having widely different chlorine contents. At low percent Cl, i.e. low extent of reaction, the spectrum is essentially that of cis-polybutadiene (15) with some additional smaller resonances due to the presence of modified butadiene units (Figure 1A). With increasing chlorine content, these additional resonances grow relative to those due to

Table I. Glass Transition Temperatures, Compositions, and Reaction Conditions of Some Dichlorocarbene Adducts of <u>Cis</u>- and <u>Trans</u>- Polybutadiene

Sample[g]	Tg (°C)[f]	Weight % Cl Chemical Analysis[a]	NMR	% RDB	Reaction Conditions[b]
1	-85	15.0	15.9	14.9	50% aqueous NaOH
2	-93	14.8	16.3	15.4	solid NaOH
3	-41	33	32.5	40.0	50% aqueous NaOH
4	-31	37.8	39.4	55.8	"
5	-5	42.1	43.4	67.2	"
6	31	47	48.0	83.5	"
7	57	51.2	51.2	97.1	solid NaOH
8	-78, 50	33.9	33.6	42.2	solid KOH
8A[c]	45	41.9	43.3	66.8	"
8B[d]	-76	22.0	22.7	23.6	"
9	-61	-	25.1	27.1	50% aqueous KOH
10	-89	17.4	19.0	18.6	solid C_sOH
11	-68	-	30.4	35.9	solid RbOH
12[e]	-56	16.0	17.4	16.7	50% aqueous NaOH
13[e]	-31, -19	31.3	31.3	37.7	"

a) Schöniger oxygen flask method.
b) see reference 14.
c) acetone-soluble fraction of sample 8.
d) acetone-insoluble fraction of sample 8.
e) Adducts of <u>trans</u>-polybutadiene.
f) Measured using a Perkin Elmer DSC-2 differential scanning calorimeter.
g) The starting polymers were >98% <u>cis</u>- or <u>trans</u>- polybutadiene.

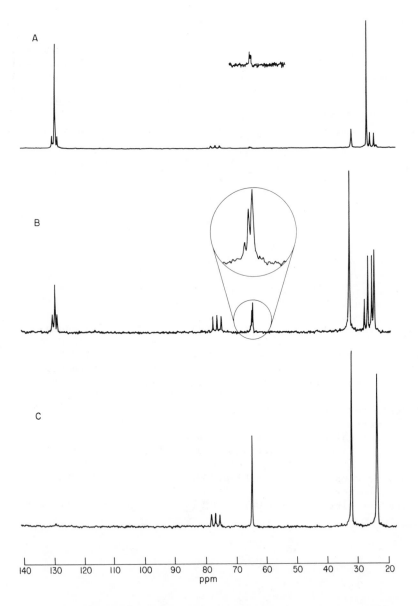

Figure 1. Carbon-13 NMR spectra of three dichlorocarbene-modified cis- polybutadienes in CDCl$_3$; (A) Sample 2, 15.4% of double bonds reacted; (B) Sample 5, 67.2% of double bonds reacted; (C) Sample 7, 97.1% of double bonds reacted. (Reproduced with permission from Ref. 16.)

the cis-polybutadiene (Figure 1B). When all double bonds have reacted, the spectrum in Figure 1C is obtained.

Peak assignments were made for the spectra in Figure 1 on the basis of model compounds (Table II), substituent effects, single-frequency off-resonance decoupling, and peak intensities. The relatively large α- and β-substitutent effects for chlorine produce ^{13}C NMR spectra of chlorinated polymers that can span the range from 25 to 100 ppm. Substituent effects on the ^{13}C shift of the central carbon are given below for chlorine substitution at various locations. Substituent effects due to vicinally or germinally substituted

$$
\begin{array}{ccc}
\pm 0.5 & +9.0 & -3.2 \\
Cl & Cl & Cl \\
\downarrow & \downarrow & \downarrow
\end{array}
$$

$$
C - C - C - \underset{\underset{Cl}{\uparrow}}{C} - C - C - C
$$

$$
+34.0
$$

chlorines are not additive, a feature which can sometimes complicate assignments for highly chlorinated polymers (8). It is interesting to note that the CCl_2 carbon in the small molecular weight models (Table II) and the polybutadiene adducts occurs about 20 ppm upfield of its usual position (5). This is due to its occurrence in a cyclopropyl ring. Our assignments for the cis-adducts are in Figure 2. A table of chemical shifts and assignments for both the cis and trans-adducts has been given elsewhere (16). The multiple resonances observed for carbons of a given functionality indicate that the ^{13}C NMR chemical shifts are sensitive to the sequence distribution of monomer units. From the changes in the intensities of the various peaks with changing extent of reaction, we have assigned peaks to copolymer triad structures such as those in Figure 3. The CCl_2 carbon is sensitive to triad structures, while the vinylene CH of the diene unit and the CH_2 carbon of the dichlorocyclopropane unit is sensitive to dyads. The CH carbon of the dichlorocyclopropane unit displays no sensitivity to sequence distribution under the present conditions.

Discussion

Isomerization during the course of the dichlorocarbene modification of polybutadienes is a potential concern. Isomerization is possible in both the unreacted butadiene portion and the chlorine-containing portion. Although the addition of dichlorocarbene to double bonds is know to be stereospecific and syn, (17) it is possible that small amounts of trans-substituted cyclopropyl groups might be present in the modified cis polymer. We saw no additional resonances suggestive of mixed stereochemistry of the cyclopropyl group in either the cis- or trans-adducts. The ^{13}C

Table II. ^{13}C NMR Chemical Shifts of Some Model Compounds

Compound	Carbon	δ(ppm)
 (cyclopropane structure with labels 4, 3, 2, 1, Cl, Cl)	1 2 3 4	67.4 25.9 19.0 20.4
 (structure with Cl, Cl, 1, 2)	1 2	72.3 31.9,32.2 (mixture of isomers)

Cl   Cl   Cl
| | |
CH - CH$_2$ - C - CH$_2$ - CH
          |
          Cl
(vinyl chloride-vinylidene
chloride copolymer) (5)
	CCl$_2$	89.7

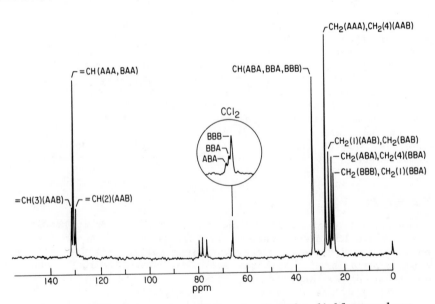

Figure 2. ^{13}C NMR peak assignments for the dichlorocarbene adducts of cis- polybutadiene.

spectrum for the modified trans-polybutadiene is sufficiently
different from that of modified cis-polybutadiene that each
structure can be detected separately in a copolymer (16). Although the ^{13}C spectrum of trans-polybutadiene homopolymer over-
laps with that of the modified cis-polybutadiene, relative peak
areas indicate that little, if any, trans-polybutadiene is present
(16). Reaction of an emulsion polybutadiene containing cis,
trans, and vinyl groups yields a ^{13}C NMR spectrum that is much
more complicated than that of either the totally cis or trans
adducts, even when the presence of vinyl groups is taken into
account.

We can obtain monomer composition and weight percent Cl from
the ^{13}C spectra using the assignments in Figure 2. The amount of
reacted double bonds, % RDB, is given by the formula

$$\% \text{ RDB} = \frac{A \text{ (CH, 32.4 ppm)}}{A \text{ (CH, 32.4 ppm)} + A \text{ (all vinylene CH)}} \times 100 \tag{1}$$

The A terms are the areas of the NMR peaks. Table I shows the
values for % RDB, as calculated using ^{13}C NMR data. Values for
weight % chlorine calculated from NMR are also given in Table I
and are compared to the corresponding values obtained by the
Schöniger method. Agreement between % chlorine by chemical
analysis and by NMR ranges from fair to excellent, depending on
sample.

A number of factors affects the accuracy and precision of
quantitative FT NMR measurements of composition. Among these are
signal-to-noise ratio, digitization of the frequency domain
spectrum, pulse repetition rate, the nuclear Overhauser effect
(NOE), and spectral resolution. The signal-to-noise ratio, the
NOE, spectral resolution, and spectrum digitization may be factors
that reduce the accuracy of the composition results. The low
signal-to-noise ratio of some of the spectra is probably a major
factor, particularly at low extent of reaction. Differential
NOE's and short pulse repetition rates can be major factors when
calculations rely on areas or intensities of both protonated and
nonprotonated carbons (18). For example, the area of the CCl_2
peak of sample 7 (Figure 1C) is only 57% of that expected, based
on the areas of the CH and CH_2 carbon peaks. The pulse repetition
rate used here will not significantly affect the accuracy of the
results in Table I since none of our quantitative results rely on
comparisons of protonated and nonprotonated carbon areas.

Number-average sequence lengths for both the dichlorocyclopropane and the cis-butadiene units can be calculated from the ^{13}C
data using the assignments in Figure 2 and standard equations.
For the chlorine-containing units (B), $\bar{n}_B$ can be obtained from
dyad concentrations using (1)

$$\bar{n}_B = \frac{N_{BB} + \frac{1}{2} N_{BA}}{\frac{1}{2} N_{BA}} = \frac{A(24.1 \text{ ppm} + 24.2 \text{ ppm}) + A(25.0 \text{ ppm})}{A(25.0 \text{ ppm})} \tag{2}$$

Here N_{BB} and N_{BA} are the concentrations of the designated dyads, where

$$N_{BB} \propto \tfrac{1}{2} A(24.1 \text{ ppm} + 24.2 \text{ ppm}) \tag{3}$$

Values for $\bar{n}_B$ can also be obtained from triad concentrations using the CCl_2 resonance ($\underline{1}$).

$$\bar{n}_B = \frac{I(65.6 \text{ ppm}) + I(65.3 \text{ ppm}) + I(65.0 \text{ ppm})}{I(65.6 \text{ ppm}) + \tfrac{1}{2} I(65.3 \text{ ppm})} \tag{4}$$

Peak intensities (I) were used for triad concentrations since they were more easily and accurately measured than the corresponding areas in this particular case. An equation (Equation 5) analogous to 2 was used for butadiene sequence lengths.

$$\bar{n}_A = \frac{A(27.4 \text{ ppm}) + A(26.3 \text{ ppm})}{A(26.3 \text{ ppm})} \tag{5}$$

The sequence lengths of both the dichlorocyclopropyl and the butadiene units for the polymers studied here are given in Table III. The agreement between $\bar{n}_B$ (dyad) and $\bar{n}_B$ (triad), although good at low sequence lengths, is poor at higher sequence lengths. The cause of this discrepancy is not known. The spectra are too simple to allow for the presence of a substantial amount of some additional structural feature that could be responsible for the discrepancy ($\underline{16}$). The use of peak heights in one case and areas in another could account for the lack of good agreement.

Another possibility is that a variation of nuclear Overhauser enhancement is occurring with composition for the peak at 24.1 ppm. This resonance is due to CH_2 carbons of both end and interior B units and the observed NOE may change as the fraction of long blocks changes. A lower NOE might be expected for carbons in a long (>3) block than in a short block on the basis of expected chain mobility ($\underline{18}$). Such an occurrence can severely complicate quantitative NMR analyses.

As expected, $\bar{n}_B$ increases and $\bar{n}_A$ decreases with increasing extent of reaction. For the samples prepared using aqueous NaOH, the observed sequence lengths are in reasonable agreement with those expected for a random copolymer displaying Bernoullian statistics (Table III) ($\underline{1}$). The random copolymer sequence lengths were calculated using Equation 6, where P_B is the mole fraction of dichlorocyclopropane units.

$$\bar{n}_A = 1/P_B \qquad\qquad \bar{n}_B = 1/(1 - P_B) \tag{6}$$

For the other polymers the $\bar{n}_B$ values are higher than calculated using Bernoullian statistics, indicating the presence of larger fractions of blocked dichlorocyclopropyl-containing units in polymers not prepared with aqueous NaOH than in comparable polymers prepared with aqueous NaOH. Comparison of samples 3 and

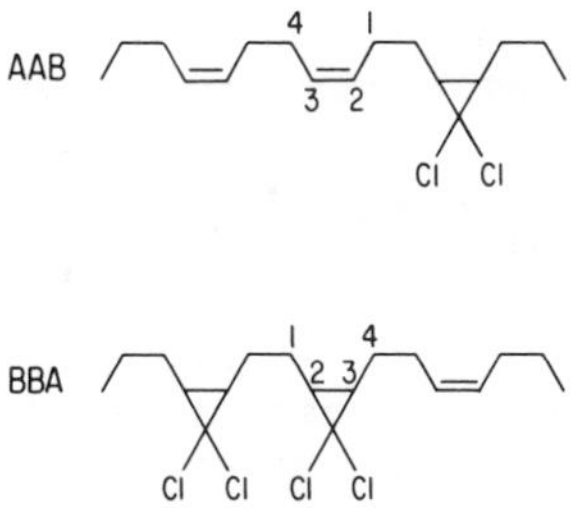

AAB

BBA

Figure 3.   Two possible comonomer triads for dichlorocarbene-modified cis-polybutadiene. (Reproduced with permission from Ref. 16.)

Table III. Number-Average Sequence Lengths of Some Dichlorocarbene Adducts of Cis-Polybutadiene

Sample	$\bar{n}_A$(dyad)	$\bar{n}_A$(theor.)b	$\bar{n}_B$(dyad)	$\bar{n}_B$(triad)	$\bar{n}_B$(theor.)b
1	5.8	6.7	1.3	1.2	1.2
2	6.8	6.5	1.5	1.8	1.2
3	2.2	2.5	1.6	1.8	1.7
4	1.8	1.8	2.1	2.4	2.3
5	1.4	1.5	2.5	3.0	3.0
6	1.3	1.2	5.3	6.6	6.1
7	-	1.0	-	-	34
8	2.9	2.4	2.5	3.3	1.7
8A	2.0	1.5	3.4	4.9	3.0
8B	3.9	4.2	1.3	1.3	1.3
9	3.2	3.7	1.3	1.5	1.4
10	6.0	5.4	1.4	1.8	1.2
11	2.8	2.8	1.6	1.8	1.6
12[a]	6.0	6.0	1.2	1.3	1.2
13[a]	2.4	2.7	1.6	1.7	1.6

a) Adducts of trans-polybutadiene.
b) Calculated using Bernoullian statistics (1).

8 well illustrates this point. Both have the same elemental composition, yet sample 8, prepared with solid KOH, has a much larger fraction of blocked units than sample 3. The difference is clearly reflected in the spectra (Figure 4). A similar comparison can be made between sample 12, an adduct of trans-polybutadiene prepared using aqueous NaOH, and sample 2, a cis-polybutadiene adduct prepared using solid NaOH.

For the samples prepared using aqueous NaOH, Tg increases in a smooth fashion with increasing extent of reaction and $\bar{n}_B$, as expected for random copolymers. The situation for samples prepared with solid base is more complex. In some cases two Tg's are observed. This has been correlated with the observation of a large domain, two-phase morphology using transmission electron microscopy (TEM) (19). In the cases where two Tg's are present, the fraction of blocked dichlorocyclopropyl units is high given the relatively low overall extent of reaction. One example is sample 8 in Tables I and III.

A question arises concerning the homogeneity in solution of the polymers with two Tg's. Can blocked segments be isolated from more random copolymer chains? And can the low and high Tg's be assigned to blocked butadiene and blocked dichloropropyl units, respectively? Sample 8, prepared with solid KOH and with Tg's of -78°C and 50°C, was fractionated using acetone. The results are in Tables I and III and Figure 5. The soluble portion (8A) had a high chlorine content, a Tg of 45°C, and a highly blocked structure [$\bar{n}_B$ (triad) = 4.9]. The insoluble portion (8B) had a low chlorine content, a Tg of -76°C, and was a random copolymer [$\bar{n}_B$ (triad) = 1.3]. This experiment links the microstructural features, as determined by ^{13}C NMR, to the thermal analysis results in a direct way. The ^{13}C NMR results for the fractionated samples agree with TEM results for other samples (14). TEM showed phase segregation of chlorine-rich and chlorine-poor regions (19). It is possible, however, that a highly heterogeneous sample like 8 could be separated into fractions with a continuous range of composition and blockiness and still display a two-phase morphology. For such heterogeneous polymers, statistical descriptions of the sequence distributions can be used to compare polymer structures, but cannot be used to draw detailed conclusions about the chemistry of preparation.

Although only a limited number of samples were prepared using alkali metal hydroxides other than NaOH, there appears to be no strong effect due to the cation in solution or in the solid.

Experimental

Details of the sample preparation are given in Reference 16. The ^{13}C NMR spectra were obtained on a Bruker HX-90E Fourier transform spectrometer at 22.6 MHz and approximately 30°C. Samples usually consisted of 0.5 g of polymer in 2 ml of $CDCl_3$ in 10 mm o.d. NMR tubes. Chemical shifts were referenced to internal $(CH_3)_4Si$. The spectral conditions were: 90° radio-frequency pulses, 15 μsec;

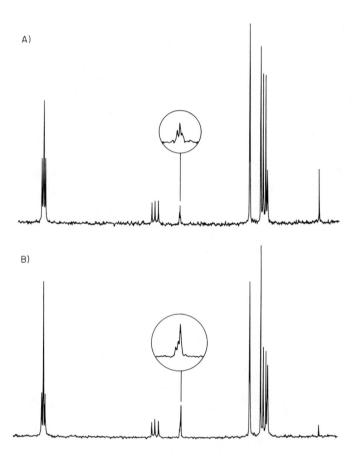

Figure 4. ^{13}C NMR spectra of two adducts with the same % Cl. (A) Sample 3, $\bar{n}_B$ (triad) = 1.8. (B) Sample 8, $\bar{n}_B$ (triad) = 3.3.

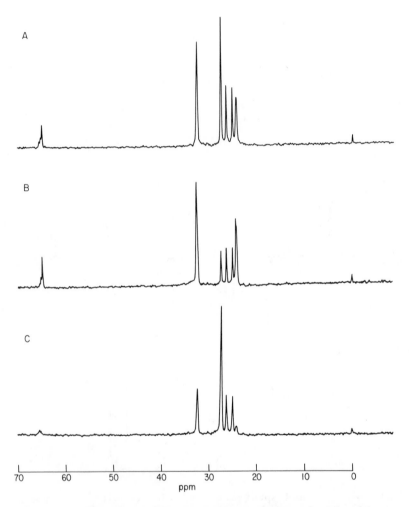

Figure 5.   Carbon-13 NMR spectra of Sample 8 (A) and its acetone soluble (8a) (B) and insoluble (8b) (C) fractions in CDCl$_3$. (Reproduced with permission from Ref. 16.)

spectral width, 6 kHz; number of points (FID), 16k; pulse repetition rate, 5 s; line broadening due to exponential filtering, 1.5 Hz; number of scans, 1000.

Acknowledgments

We thank Dr. C. J. Singleton for providing the Tg results, R. W. Smith for providing the TEM results, and R. E. Scourfield for acquiring the ^{13}C NMR spectra.

Literature Cited

1. Randall, J. C. "Polymer Sequence Determination", Academic: New York, 1977.
2. Keller, F.; Zepnik, S.; Hösselbarth, B. Faserforsch. Tertiltech. 1978, 29, 490.
3. Keller, F.; Mügge, C. Faserforsch. Textiltech. 1976, 27, 347.
4. Carman, C. J. ACS Symp. Ser. 1980, 142, 81.
5. Komoroski, R. A.; Shockcor, J. P. Macromolecules 1983, 16, xx.
6. Kaplan, S. J. Polym. Sci., Polym. Chem. Ed. 1980, 18, 3307.
7. Keller, F.; Hösselbarth, B. Faserforsch. Textiltech. 1978, 29, 152.
8. Komoroski, R. A.; Parker, R. G.; Lehr, M. H. Macromolecules 1982 15, 844.
9. Keller, F.; Mügge, C. Plaste u. Kautschuk 1977, 24, 88.
10. Pinazzi, C.; Levesque, G. C. R. Acad. Sci. 1965, 260, 3393.
11. Lishanskii, I. S.; Tsitskhtsev, V. A.; Vinogradova, N. D. Vysokom. Soedin. 1966, 8, 186.
12. Lal, J.; Saltman, W. J. Poly. Sci. A-1 1966, 4, 1637.
13. DeWitt, W.; Hurwitz, M. J.; Albright, F. J. Poly. Sci. A-1 1969, 7, 2453.
14. Horne, S. E., Jr., to be published.
15. Clague, A. D. H.; van Broekhoven, J. A. M.; Blaauw, L. P. Macromolecules 1974, 7, 348.
16. Komoroski, R. A.; Horne, S. E., Jr.; Carman, C. J. J. Polym. Sci., Polym. Chem. Ed. 1983, 21, 89.
17. Morrison, R. T.; Boyd, R. N. "Organic Chemistry", Allyn and Bacon: Boston, 1973; p. 311.
18. Komoroski, R. A. J. Polym. Sci., Polym. Phys. Ed. 1979, 17, 45.
19. Horne, S. E., Jr.; Singleton, C. J.; Smith, R. W., unpublished results.

RECEIVED September 22, 1983

NMR Spectra of Styrene Oligomers and Polymers

HISAYA SATO and YASUYUKI TANAKA

Faculty of Technology, Tokyo University of Agriculture and Technology, Koganei, Tokyo 184, Japan

Styrene oligomers having propyl end groups were prepared by the oligomerization of styrene initiated with ethyllithium and terminated with 1-bromopropane. The oligomers were separated according to molecular weight by GPC and the 2-5 mers were fractionated into diastereomers by liquid chromatography. The diastereomers of the 2- and 3-mers were identified by 1H NMR and those of the 4- and 5-mers by using the results of the 2- and 3-mers. The 13C NMR signals of the oligomers were assigned by selective decoupling measurements and also by comparing the chemical shifts with each other. The methylene and phenyl C(1) signals of polystyrene were assigned on the basis of the oligomer. Polystyrenes prepared with benzoyl peroxide, n-butyllithium, and trifluoroboron etherate had a random distribution of r and m dyads with P_r values of 0.54, 0.56, and 0.45, respectively.

The 13C NMR spectrum of polystyrene was first reported by Bovey et al. in 1970 ($\underline{1}$). The methylene and phenyl C(1) carbon resonances display splittings which reflect the tacticity distribution and have been assigned to configurational sequences by many authors ($\underline{2-5}$). However, the assignments differed from each other and different values, ranging from 0.45 to 0.72, were reported for the probability of a racemic dyad (P_r) in radically prepared polystyrene. These differences may arise from the fact that the only polystyrene with a known structure is isotactic polystyrene prepared with a Ziegler-Natta catalyst. Using this polymer one can only assign resonances due to the mmm tetrad and the $mmmm$ pentad, while other signals are arbitrarily assigned by assuming a Bernoullian or other statistical distribution.

In order to assign resonances in the 1H and 13C NMR spectra of polystyrene, styrene oligomers having methyl end groups have been prepared and separated into diastereomers up to tetramer

0097-6156/84/0247-0181$06.00/0

(6–9). However, these oligomers failed to provide useful information concerning the spectral analysis of polystyrene, because the
methyl end groups resulted in special chemical shifts for the 1H
and 13C atoms located nearby.   Apparently, these oligomers had
different conformational distributions than the polymer.

   In this chapter 1H and 13C NMR spectra of polystyrene are
analysed utilizing styrene oligomers having propyl end groups.

Preparation and Separation of Styrene Oligomers

Styrene oligomers having propyl end groups were prepared by
initiating oligomerization with ethyllithium and terminating with
1-bromopropane in ethyl ether:

$$CH_3CH_2Li + CH_2{=}CH_{Ph} \longrightarrow CH_3CH_2(CH_2CH_{Ph})_nLi$$

$$\xrightarrow{CH_3CH_2CH_2Br} CH_3CH_2(CH_2CH_{Ph})_nCH_2CH_2CH_3 \qquad (1)$$

The average degrees of oligomerization were controlled by varying
the molar ratio of styrene and the initiator.   This anionic
oligomerization method is much easier than the Grignard and other
coupling methods used for the preparation of oligomers with methyl
end groups, although this product is a mixture of oligomers having
different degrees of oligomerization.

   In order to separate the oligomer mixture, we examined the
elution behavior of the oligomer using a styrene–divinyl benzene
copolymer gel as a stationary phase and several eluents having
different polarity as shown in Figure 1.   Eluents having the same
polarity as the gel (solubility parameter(SP)= ca. 9.1), such as
chloroform and tetrahydrofuran caused the oligomer to separate
easily from higher molecular weight polymer.   The oligomers eluted
much earlier when these solvents were used than when other types
of eluents were used.   This indicates that the oligomers are
separated by GPC by a size exclusion mechanism.

   Nonpolar eluents (2,2,4-trimethylpentane and isopropyl ether)
and polar eluents (methanol and acetonitrile) eluted the oligomers
more slowly than chloroform by a factor of two.   It is clear that
these eluents cause separation to be controlled by an adsorption
mechanism with a negligible size exclusion effect.   Using these
eluents, we observed peak separation or broadening due to the
diastereomers for the dimer and higher oligomers.

   Using cyclohexane and acetone, which have similar SP values
to that of the gel, we separated the oligomers into two groups;
one at an elution volume around 20 ml due to higher molecular
weight oligomers and the other around 30 – 40 ml due to lower
molecular weight ones.   This discontinuous molecular weight-
elution volume relation can be explained by a hybrid mechanism of
size exclusion and adsorption effects.   Larger oligomers cannot
permeate into the gel and consequently elute earlier, experiencing

only a slight adorption effect, while smaller oligomers permeate into the gel and suffer both size exclusion and adsorption effects. They elute at almost the same elution volume regardless of the molecular weight.

The oligomer mixture prepared according to Equation (1) was first separated into pure n-mers by GPC using chloroform as an eluent. The pure 2 - 5 mers were separated by adsorption chromatography using a recycling technique. Diisopropyl ether was used as an eluent, because this eluent provided a column with a higher number of theoretical plates than did trimethylpentane, acetonitrile, and methanol. The dimers and trimers were separated into 2 and 3 fractions, respectively, after 12 recycles (Fig. 2 a and b). The tetramer, which has 6 diastereomers, was separated accordingly into 6 fractions after 24 recycles, although the separation of the second and the third fractions was not so good (Fig. 2 c). The separation of the pentamer was carried out in two stages of fractionation. In the first stage, the first and the ninth fractions were collected and the remaining portion was separated into 3 parts labelled A, B, and C. The three parts were subsequently separated into fractions 2 and 3, fractions 4, 5, and 6, and fractions 7 and 8, respectively, during the second stage of fractionation (Fig. 2 d).

NMR Spectra of the Dimer and Trimer

The diastereomers of the dimer were identified through the splitting pattern observed for the methylene protons flanked by two methine protons; the methylene protons of the r dyad are equivalent, while those of the m dyad are nonequivalent.

$$CH_3CH_2CH_2-\overset{\overset{\displaystyle Ph}{|}}{\underset{\underset{\displaystyle H}{|}}{C}} - \overset{\overset{\displaystyle \mathbf{Ha}}{|}}{\underset{\underset{\displaystyle \mathbf{Ha}}{|}}{C}} - \overset{\overset{\displaystyle H}{|}}{\underset{\underset{\displaystyle Ph}{|}}{C}}-CH_2CH_2CH_3 \qquad CH_3CH_2CH_2-\overset{\overset{\displaystyle H}{|}}{\underset{\underset{\displaystyle Ph}{|}}{C}} - \overset{\overset{\displaystyle \mathbf{Ha}}{|}}{\underset{\underset{\displaystyle \mathbf{Hb}}{|}}{C}} - \overset{\overset{\displaystyle H}{|}}{\underset{\underset{\displaystyle Ph}{|}}{C}}-CH_2CH_2CH_3$$

r isomer m isomer

Figure 3 shows methine and methylene H-1 NMR spectra of the 2 dimer fractions which gave methylene proton resonances around 1.8 ppm. Fraction 2-1 showed an $AA'X_2$ methylene resonance pattern, while fraction 2-2 showed an ABX_2 system. The methylene proton signals decoupled from the methine proton display more clearly the difference between the splitting patterns; Fr. 2-1 exhibited AA' splittings, while Fr. 2-2 exhibited an AB quartet. Therefore, fractions 2-1 and 2-2 were assigned to the r and m isomers, respectively. Similarly, the trimer fractions, Fr. 3-1, 3-2, and 3-3, were identified as rr, rm, and mm isomers by the splitting pattern of their methylene proton signals as shown in Figure 4.

The methine protons of the dimer and the trimer resonated in two chemical shift ranges. One occurred between 2.5-2.2 ppm, which is assigned to the m end and the mm center and the other between 2.2-2.0 ppm, which is assigned to the r end and the (rr + rm) center. It is noteworthy that the central methylene proton of

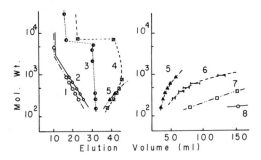

Figure 1. Relationship between molecular weight and elution volume of styrene oligomers in chloroforn (curve 1; SP= 9.1), tetrahydrofuran (2; 9.1), cyclohexane (3; 8.2), acetone (4; 9.4), diisopropyl ether (5; 7.3), 2,2,4-trimethylpentane (6; 7.0), acetonitrile (7; 11.8), methanol (8; 12.9). (column: 10 mm i.d. x 50 cm). Reproduced with permission from Ref. 12 Copyright 1981, Hüthig & Wepf Verlag.

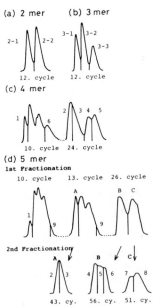

Figure 2. Separation of styrene oligomers into diastereomers. (Column: 21 mm i.d. x 60 cm x 3; Eluent: diisopropyl ether)

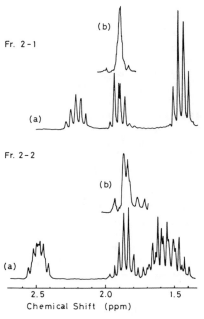

Fr. 2-1

(b)

(a)

Fr. 2-2

(b)

(a)

2.5   2.0   1.5
Chemical Shift (ppm)

Figure   3. H-1 NMR spectra of the dimer (at 200 MHz):
non-decoupled spectra (a) and methylene spectra decoupled
from methine protons (b).

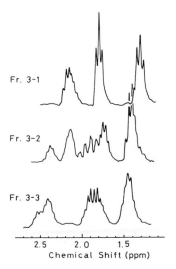

Fr. 3-1

Fr. 3-2

Fr. 3-3

2.5   2.0   1.5
Chemical Shift (ppm)

Figure 4.   H-1 NMR spectra of the
trimer (at 200 mHz).

the rm isomer resonated at the same frequency as that of the rr isomer and higher than that of the mm isomer. This finding is consistent with the result of Shepherd et al. (10), who reported that the methine proton resonance of polystyrene is split into two peaks, the one at the lower magnetic field is assigned to the mm triad, while the other is assigned to the rm and rr triads. However, in the case of the styrene trimer with methyl end groups, the central methine protons of mm, rm, and rr isomers were observed at chemical shifts of 2.5, 2.3, and 2.1 ppm, respectively (8). This discrepancy may be explained by the small end groups, which may provide the oligomer a special conformation or render a specific chemical shift to the proton.

The styrene dimer with propyl end groups has the end methyl (A), end methylene (B), external methylene (C), end methine (D) and inner methylene carbons (E) in the main chain and the end phenyl group (i) as the side chain. The trimer has one new type of carbon atom, i.e., inner methine (D'), and the inner phenyl group (ii) along with the carbons and the phenyl group contained in the dimer. Furthermore, the tetramer possesses a central methylene carbon (F), and the pentamer possesses a central methine (D") and a central phenyl group (iii).

2 mer:
$$\underset{\underset{Ph_{(i)}}{|}}{A \quad B \quad C \quad D \quad E \quad D \quad C \quad B \quad A}$$
$$CH_3-CH_2-CH_2-\underset{Ph_{(i)}}{CH}-CH_2-\underset{Ph_{(i)}}{CH}-CH_2-CH_2-CH_3$$

3 mer:
$$A \quad B \quad C \quad D \quad E \quad D' \quad E \quad D \quad C \quad B \quad A$$
$$CH_3-CH_2-CH_2-\underset{Ph_{(i)}}{CH}-CH_2-\underset{Ph_{(ii)}}{CH}-CH_2-\underset{Ph_{(i)}}{CH}-CH_2-CH_2-CH_3$$

4 mer:
$$A \quad B \quad C \quad D \quad E \quad D' \quad F \quad D' \quad E \quad D \quad C \quad B \quad A$$
$$CH_3-CH_2-CH_2-\underset{Ph_{(i)}}{CH}-CH_2-\underset{Ph_{(ii)}}{CH}-CH_2-\underset{Ph_{(ii)}}{CH}-CH_2-\underset{Ph_{(i)}}{CH}-CH_2-CH_2-CH_3$$

5 mer:
$$A \quad B \quad C \quad D \quad E \quad D' \quad F \quad D'' \quad F \quad D' \quad E \quad D \quad C \quad B \quad A$$
$$CH_3-CH_2-CH_2-\underset{Ph_{(i)}}{CH}-CH_2-\underset{Ph_{(ii)}}{CH}-CH_2-\underset{Ph_{(iii)}}{CH}-CH_2-\underset{Ph_{(ii)}}{CH}-CH_2-\underset{Ph_{(i)}}{CH}-CH_2-CH_2-CH_3$$

Carbons A, B, C, D + D', and E were assigned by selective decoupling measurements. For example, the rm isomer of the trimer showed 7 signals in the region of 46 – 38 ppm (Fig. 5 a). By decoupling the protons on carbon C at 1.45 ppm signals 6 and 7 appeared as sharp peaks (Fig. 5 b). Therefore, the two signals were assigned to the carbon C. After comparison with the other isomers, signal 6 was assigned to carbon C on the r side and signal 7 to the m side. Decoupling the protons on carbons E in the r and m sides at 1.75 and 1.92 ppm, respectively, led to sharp lines for resonances 1 and 4 (Fig. 5 c and d), showing that these signals are due to carbons E of the r and m sides. Upon decoupling the methine protons, we were able to assign resonances

2, 3, and 5 to the methine carbons D and D'. Resonances 2 and 3 were assigned to carbon D and resonace 5 to D' by comparison with the other isomers and the dimer.

In the dimer and trimer, carbons A and B resonated at 14 and 20 ppm, respectively, regardless of the molecular weight and configuration. Carbon C resonated around 40 ppm for the r side and 38 ppm for the m side. Carbons D and D' resonated at 43 ppm and 41 ppm regardless of configuration. Carbon E resonated in a wide chemical shift range of 46 to 40 ppm, reflecting a sensitivity to configurational sequences.

Each isomer of the trimer has two or three types of phenyl C(1) carbons. The phenyl C(1) resonaces of the rr and mm isomers were assigned by the signal intensity ratio. Those of the rm isomer were assigned through comparison of these chemcial shifts with other isomers. The phenyl C(1) carbon of the end unit resonated from lower magnetic field; $\underline{mm} < \underline{mr} < \underline{rr} < \underline{rm}$. The inner unit gave an order ; $\underline{mm} < \underline{rm} < \underline{rr}$. This order also held for the phenyl C(1) carbon of the tetramer and the pentamer.

Identification of the Tetramer and Pentamer

The diastereomers of the tetramer and the pentamer are difficult to identify using only the splitting pattern of the methylene protons. Several methylene proton resonances overlap around 1.8 ppm, and the methylene protons adjacent to the m dyad (eg. the $\underline{rm}$ proton in the rm isomer of the trimer) display an AB type splitting pattern in the dimer and the trimer. Therefore, the isomers of the tetramer and the pentamer were identified by using the following information concerning the dimer and trimer:
1) the elution order of liquid chromatography:
   the more r dyads an isomer has, the earlier it elutes
   (eluting order was $r < m$ for the dimer and $rr < rm < mm$ for the trimer.)
2) chemical shift range of methine proton:
   2.5–2.2 ppm for the m end and the mm center
   2.2–2.0 ppm for the r end and the (rm + rr) center.
3) chemical shift of external carbon (C):
   ca. 40 ppm for r side and ca. 38 ppm for the m side.

The composition of the diastereomers in each fraction of the tetramer and the pentamer was determined by the signal intensity of the C carbon as listed in Table I. After referring to Figure 2, it is clear that the tetramer can be divided into 4 groups having 3, 2, 1 and 0 r dyads. Any further separations of diastereomers with the same number of r dyads are too difficult. Similarly, the pentamer was first divided by the number of r dyads and then separated according to diastereomers having the same number of r dyads, i.e., two diastereomers of r_3m (part A in Fig. 2), four diastereomers of r_2m_2 (part B), and two diastereomers of rm_3 (part C). These separations were difficult and were conducted through a second stage of fractionation.

Table I.   Composition of 4 and 5 Mers

4 mer		5 mer			
Fraction	Composition	Fraction	Composition	Fraction	Composition
1	*rrr* 95%	1	*rrrr* 95%	5	*rrmm* 50%
2	*rrm* 90%	2	*rrmr* 67%		*rmmr* 11%
3	*rrm* 83%		*rrrm* 33%		*mrrm* 32%
	rmr 17%	3	*rrrm* 26%	6	*mrrm* 46%
4	*mmr* 90%		*rrrr* 74%		*mrmr* 54%
5	*mrm* 85%	4	*rrrm* 84%	7	*mmmr* 95%
6	*mmm* 90%		*rmmr* 16%	8	*mmrm* 83%
				9	*mmmm* 83%

13C NMR Spectra of the Tetramer and Pentamer

The aliphatic carbon signals of the tetramer and the pentamer were
assigned by selecive decoupling measurements to carbons A, B, C,
D, E + F, D' + D". In the tetramer and pentamer resonances for
carbons A, B, C, D and D' had the same chemical shifts as observed
for the dimer and trimer. The resonances for carbon D" occurred
at 41 ppm, which is the same as carbon D', regardless of configuration. Depending on the configurational sequences, carbon F
resonated over a wide chemical shift range of 47 to 41 ppm, which
is almost equal to the chemical shift range observed for carbon E.
The resonances from carbons E and F were assigned by the signal
intensity ratio in the case of symmetric diastereomers of the
tetramer, i.e., *rrr*, *rmr*, *mrm*, and *mmm* isomers. In the case of
the other isomers, assignments were made by comparison of the
observed chemical shifts to those of the corresponding carbons in
the dimer and trimer as shown in Figure 6. Good correlations are
observed among the chemical shifts of the carbons corresponding to
E and F, which indicates the validity of the identification of the
diastereomers.

The phenyl C(1) signals of the end, the inner and the central
units were assigned by comparison of the chemical shifts of the
corresponding carbons of the isomers as shown in Figure 7. The
phenyl C(1) carbon signal of the end unit appeared from lower
magnetic field; *mm* < *mr* < *rr* < *rm*. The resonating order of the
phenyl C(1) signal of the inner unit was *mm* < *rm*(*mr*) < *rr*. In the
case of C(1) signal of the central unit in the pentamer the resonating order was also *mm* < *mr*(*rm*) < *rr*, except for the *rmmr* signal,
which resonated in the middle of signals due to *mr* sequences.

The C(1) signal of the inner unit of the methyl end oligomer
has been reported by Jasse et al, with the resonating order from
lower magnetic field being *rrm* ≃ *mrr* < *rmm* (or *mmr*) < *rmr* < *mmm* < *mrm*
(or *rmm*) ≃ *rrr* < *mrm* (9). Thus, the methyl end oligomer showed no
regularity in the resonating order of the inner C(1) signal at the
triad sequence level. In view of fact that the phenyl C(1) signal

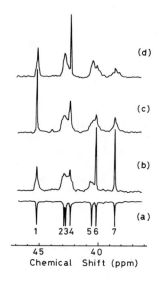

Figure 5.   C-13 NMR spectra of the rm isomer of the trimer: complete H-1 decoupled (a), selective H-1 decoupled at 1.45 ppm (b), 1.75 ppm (c), and 1.92 ppm (d).

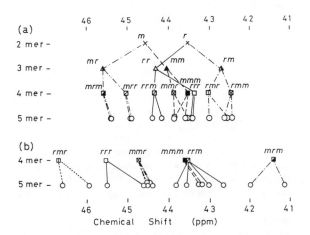

Figure 6. Chemical shift of the internal (a) and the central methylene carbon (b).

of the mm triad has been assigned to the lowest magnetic field by comparison with the spectrum of isotactic polystyrene, it is peculiar that the mmm isomer of the methyl end oligomer gave an inner phenyl C(1) resonance in the middle of those for the other isomers. Therefore, the methyl end oligomer can not be a good model compound for analyzing the phenyl C(1) resonances of polystyrene.

13C NMR Signal Assignment of Polystyrene

As shown in Figure 6 the central methylene carbon of the tetramer and the pentamer resonated from lower magnetic field; $rmr < rrr <$ $rmm < rrm + mmm < mrm$. In the pentamer the central methylene carbon in the mmr and mmm tetrads resonated over a small chemical shift range, while that in the rmr, rrr, rrm, and mrm tetrads resonated over a wide chemical shift range. Therefore, it is presumed that the methylene carbon of polystyrene resonated from lower magnetic field; $rmr < rrr < rmm < rrm + mmm < mrm$ and that the signals due to the rmr, rrr, rrm, and mrm tetrads show splittings due to the hexad configuration.

Figure 8 shows 13C NMR spectra of polystyrene prepared with benzoyl peroxide catalyst and measured in chloroform-d at room temperature (a) and in 1,2-dichlorobenzene at 150°C (b). The spectrum measured at room temperature was assigned as shown in the figure on the basis of the results from the oligomer.

The spectrum measured at 150°C showed well resolved signals, which were also assigned on the basis of results from the oligomer. Signals 1, 2, and 3 were attributed to the rmr tetrads, which were assigned to the $mrmrm$, $mrmrr$, and $rrmrr$ from the lower magnetic field, because in the case of the styrene pentamer the methylene carbon of the $rmrm$ isomer resonated at lower magnetic field than that of the $rmrr$ isomer. Similarly, signals 4, 5, and 6 were assigned to the hexad configurations having the rrr tetrad, and signals 10, 11, 12 to the hexads having the mrm tetrad. Signal 8 and part of signal 9 were allotted to the rrm tetrad and were assigned to the hexad sequences considering the signal intensities. Signals due to the mmr and mmm tetrads (signals 7 and 9) showed no discrete splitting due to hexad configurations. This signal assignment is almost the same as one reported by Chen et al. (11).

The intensity of each signal of the spectrum measured at 150°C was determined for polystyrenes prepared with benzoyl peroxide, n-butyllithium, and trifluoroboron etherate catalysts. The observed relative intensities of the signals were in good agreement with the calculated values assuming Bernoullian statistics with P_r of 0.54, 0.56, and 0.45 for the radical, anionic, and cationic polystyrenes, respectively (Table II). The P_r value of the radical polystyrene was nearly the same as that reported by Shepherd et al. (10).

The phenyl C(1) carbon resonance of polystyrene was reported

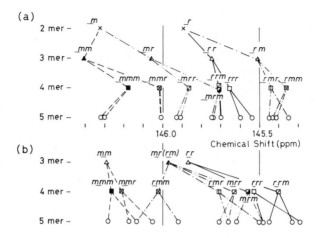

Figure 7. Chemical shift of the phenyl C(1) carbon in the end (a) and inner units (b).

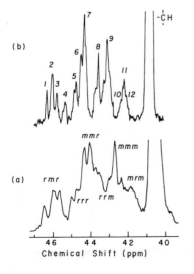

Figure 8. Methylene carbon spectra of radical polystyrene measured in chloroform-d at room temperature (a) and in 1,2-dichlorobenzene at 150 °C (b). Reproduced with permission from Ref. 13. Copyright 1983, John Wiley & Sons, Inc.

Table II.   Assignment of Methylene Carbon Signal
            (Measured in 1,2-Dichlorobenzene at 150°C)

Signal	Chemical shift (ppm)	Assignment	Intensity (%)					
			BPO		BuLi		BF$_3$·Et$_2$O	
			obsd.	(Pr=0.54)	obsd.	(Pr=0.56)	obsd.	(Pr=0.45)
1	46.73	*mrmrm*	2.6	2.8	2.4	2.7	3.8	3.4
2	46.44	*mrmrr*	6.4	6.1	6.1	6.8	5.5	5.5
3	46.22	*rrmrr*	4.0	3.9	4.5	4.3	2.6	2.3
4	45.61	*mrrrm*	3.6	3.4	4.0	3.4	2.9	2.8
5	45.22	*rrrrm*	7.6	7.8	10.1	8.7	4.7	4.5
6	44.95	*rrrrr*	4.3	4.6	5.4	5.5		
7	44.79	*mmr*	21.5	22.9	21.5	21.7	29.4	29.0
8	44.05	*mrrmr* *mrrmm*	12.8	12.4	11.3	12.1	12.5	12.2
9	43.59	*rrrmr* *rrrmm* *mmm*	23.9	24.2	24.6	24.0	24.4	26.7
10	42.70	*rmrmr*	2.7	3.3	2.4	3.4	2.6	2.8
11	42.66	*mmrmr*	6.7	5.7	5.3	5.3	6.5	5.7
12	42.55	*mmrmm*	2.9	2.4	2.3	2.1	5.1	4.1

to show splittings due to pentad configurations ($\underline{1}$, $\underline{2}$, $\underline{3}$, $\underline{5}$). Therefore, the pentamer is required as the smallest model compound to analyze these resonances. Figure 9 shows the chemical shift behavior of the phenyl C(1) carbon in the central unit of the pentamer as measured at room temperature in chloroform-d. The central phenyl C(1) carbon signal appeared in order of increasing magnetic field as $mm < rm < rr$ triads.

Figure 10 shows the phenyl C(1) spectra of the radical polystyrene measured at room temperature in chloroform-d (a) and at 150°C in 1,2-dichlorobenzene (b). The resonances observed at room temperature were assigned as shown in Figure 10 on the basis of the resonating order of the pentamer. The spectrum measured at 150°C was also assigned on the basis of the pentamer as shown in Table III. The resonating order is the same as that for the pentamer except for the signal from the $rmmr$ pentad. The relative intensities of the peaks were measured for the polystyrenes prepared with benzoyl peroxide, n-butyllithium, and trifluoroboron etherate initiators (Table III). The observed relative intensities for the three types of polystyrenes were in good agreement with values calculated assuming Bernoullian statistics. The P_r values used in the Bernoullian calculation were determined from the methylene carbon resonances. This means that the same configurational sequences, which supports the validity of the signal assignments, could be determined from either the methylene or phenyl C(1) resonances for these polystyrenes.

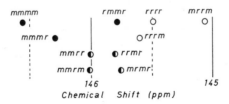

Figure 9. Chemical shift of the central phenyl C(1) carbon of the pentamer.

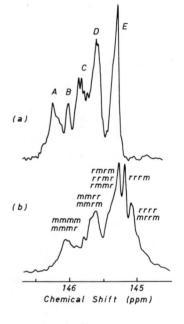

Figure 10. Phenyl C(1) carbon spectra
of radical polystyrene measured in
chloroform-d at room temperature (a)
and in 1,2-dichlorobenzene at 150 °C
(b). Reproduced with permission from
Ref. 13. Copyright 1983, John Wiley &
Sons, Inc.

Table III. Assignment of Phenyl C(1) Carbon Signal
(Measured in 1,2-Dichlorobenzene at 150°C)

Signal	Chemical Shift (ppm)	Assignment	Intensity (%)					
			BPO obsd.	(Pr=0.54)	BuLi obsd.	(Pr=0.56)	BF$_3$·Et$_2$O obsd.[3]	(Pr=0.45)
A	146.23	*mmmm* *mmmr*	16.3	15.0	14.2	13.3	22.5	24.1
B	146.03	*rmmr*	7.5	6.2	6.6	6.1	6.9	6.1
C	145.80	*mmrm* *mmrr*	21.8	22.8	20.3	21.7	28.1	27.2
D	145.63	*rmrm* *rrmr*	27.6	26.8	27.4	27.6	21.8	22.3
E	145.32	*rr*	26.8	29.2	31.5	31.3	20.7	20.3

Source: Ref. 13. Used by permission of John Wiley & Sons, Inc. Copyright 1983.

Literature Cited

1. L. F. Johnson, F. Heatley, and F. A. Bovey, Macromolecules, 3, 175 (1970).
2. Y. Inoue, A. Nishioka, R. Chûjo, Makromol. Chem., 156, 207 (1972).
3. K. Matsuzaki, T. Uryu, K. Osada, T. Kawamura, Macromolecules, 5, 816 (1972).
4. J. C. Randall, J. Polym. Sci., Polym. Phys. Ed., 13, 889 (1975).
5. S. Sparno, J. Lacoste, S. Raynal, J. F. Regnier, F. Schué, R. Sempere, J. Sledz, Polymer J., 12, 861 (1980).
6. F. A. Bovey, F. P. Hood III, E. W. Anderson, L. C. Snyder, J. Chem. Phys., 42, 3900 (1965).
7. D. Doskocilova, B. Schneider, J. Polym. Sci., Part B 3, 213 (1965).
8. D. Lím, M. Kolínský, J. Petránek, D. Doskocilová, B. Schneider, J. Polym. Sci., Part B 4, 645 (1966).
9. B. Jasse, F. Lauprêtre, L. Monnerie, Makromol. Chem., 178, 1987 (1977).
10. L. Shepherd. T. K. Chen, H. J. Harwood, Polym. Bull., 1, 445 (1979).
11. T. K. Chen, T. A. Gerken, H. J. Harwood, Polym. Bull., 2, 37 (1980).
12. H. Sato, K. Saito, K. Miyashita, Y. Tanaka, Makromol. Chem., 182, 2259 (1981).
13. H. Sato, Y. Tanaka, K. Hatada, J. Polym. Sci., Polym. Phys. Ed., 21 (1983).

RECEIVED November 3, 1983

75-MHz ^{13}C NMR Studies on Polystyrene and Epimerized Isotactic Polystyrenes

H. JAMES HARWOOD, TENG-KO CHEN, and FU-TYAN LIN[/]

Institute of Polymer Science, The University of Akron, Akron, OH 44325

Isotactic polystyrene was epimerized to various extents by reaction with KOtBu in hexamethylphosphoramide solution at 100°. The ^{13}C-NMR spectra of the epimerized samples and of polystyrene in 9:1 trichlorobenzene:nitrobenzene-d$_6$ solution at 150° were recorded and analyzed using stereosequence distributions that were calculated for the samples by Monte Carlo simulation of the epimerization process. Triad, pentad and heptad assignments were developed for the aromatic C-1 carbon resonance patterns and triad assignments previously proposed for the methine carbon resonances of polystyrene were verified. Based on the assignments developed, the ^{13}C-NMR spectrum of polystyrene indicates it to be almost atactic, P(m)=0.45.

Polystyrene was one of the first polymers to be studied by high resolution NMR spectroscopy(1), but its spectra are incompletely understood even today. The ^{1}H-NMR spectra of polystyrene are poorly defined; the methine and ortho aromatic proton spectra are influenced significantly by pentad or larger stereosequence effects but they contain only three or four broad, badly overlapped resonances. In contrast, the ^{13}C-NMR spectra of polystyrene are rich in detail, being sensitive to hexad and heptad stereosequence effects. Since there are 20 distinguishable hexad combinations and 36 distinguishable heptad combinations, it has been extremely difficult to assign the various resonances observed in the ^{13}C-NMR spectrum of polystyrene. The difficulty has been compounded by the fact that free radical initiated polymerizations of styrene yield polymer having the same microstructure, regardless of polymerization temperature. Many ionic polymerizations of styrene also yield polymers that have the same microstructure as those prepared

[/]Current address: Department of Chemistry, University of Pittsburgh, Pittsburgh, PA 15261.

by free radical polymerization. Thus, it has not been possible to prepare, by direct polymerization, a series of polystyrene samples having microstructures that varied in some rational way. Studies on the spectra of such samples, in which the intensities of the various signals would vary as the polymer structure changed, would have been very useful for making resonance assignments. Cationic and some anionic polymerizations of styrene have yielded polymers that have spectra that differ from the spectrum of free radical polymerized polystyrene(2-3). However, the usefulness of such samples was dampened by the possibility that side reactions accompanied the cationic polymerizations and by the possibility that multi-state propagation mechanisms were involved in some of the anionic polymerization reactions.

Our recent understanding of the NMR spectra and structure of polystyrene is based on the results of model compound studies(4-11), theoretical calculations(12-14), studies on polystyrene derivatives (15) and analogues(16), studies on the NMR spectra of epimerized isotactic polystyrene samples(17-19) and on results obtained using other approaches(20-22). The model compound approach, exemplified by the excellent paper contained in this volume(11), involved isolating polystyrene oligomers of known configuration and recording their NMR spectra. Chemical shift assignments made for the oligomers are then used in making assignments for polystyrene. This approach requires that a large number of different oligomers be separated and studied if assignments are to be made at pentad or higher levels and care must be taken to design the oligomers so that they adopt conformations similar to those adopted by the polymer. This tedious approach has been very fruitful and rewarding(11).

Epimerization of stereoregular polymers(17-19,23-33) provides a way to obtain polymers having controlled sequence distributions. The products of polymer epimerization reactions can be useful for establishing NMR assignments because their structures can be calculated by simulating the epimerization process. They can thus serve as models for structure assignment purposes. Although free radical chemistry has been used for some polymer epimerization experiments(30), strong bases are generally used to epimerize polymers. Shepherd(34), made a major contribution to the study of polystyrene structure by showing that isotactic polystyrene could be epimerized by potassium t-butoxide in hexamethylphosphoramide solution at mild temperatures. This process can be envisioned as shown by the equations given on the following page.

These equations show how a mmmm sequence in polystyrene can be converted into a mrrm sequence by a simple epimerization event. Should the configuration of the fourth carbon from the left in the last structure also be altered, a rmrm pentad would result. Thus, by a series of epimerization steps it is possible to change isotactic polystyrene gradually into a polymer that exhibits the same NMR spectra as the polystyrene that was prepared by free radical polymerization(17-19). A Monte Carlo program has

$$\begin{array}{cccccccccc}
& H & & H & & H & & H & & H \\
& | & & | & & | & & | & & | \\
\sim CH_2 & -C & -CH_2 & -C & -CH_2 & -C & -CH_2 & -C & -CH_2 & -C\sim \\
& | & & | & & | & & | & & | \\
& \phi & & \phi & & \phi & & \phi & & \phi \\
& & m & & m & & m & & m &
\end{array}$$

$$tBuO^{\ominus} \quad \downarrow\uparrow \quad tBuOH$$

$$\begin{array}{cccccccccc}
& H & & H & & & & H & & H \\
& | & & | & & \ominus & & | & & | \\
\sim CH_2 & -C & -CH_2 & -C & -CH_2 & -C & -CH_2 & -C & -CH_2 & -C\sim \\
& | & & | & & | & & | & & | \\
& \phi & & \phi & & \phi & & \phi & & \phi
\end{array}$$

$$\downarrow\uparrow$$

$$\begin{array}{cccccccccc}
& H & & H & & \phi & & H & & H \\
& | & & | & & | & & | & & | \\
\sim CH_2 & -C & -CH_2 & -C & -CH_2 & -C & -CH_2 & -C & -CH_2 & -C\sim \\
& | & & | & & \ominus & & | & & | \\
& \phi & & \phi & & & & \phi & & \phi
\end{array}$$

$$tBuOH \quad \downarrow\uparrow \quad tBuO^{\ominus}$$

$$\begin{array}{cccccccccc}
& H & & H & & \phi & & H & & H \\
& | & & | & & | & & | & & | \\
\sim CH_2 & -C & -CH_2 & -C & -CH_2 & -C & -CH_2 & -C & -CH_2 & -C\sim \\
& | & & | & & | & & | & & | \\
& \phi & & \phi & & H & & \phi & & \phi \\
& & m & & r & & r & & m &
\end{array}$$

been written to simulate this epimerization process and to calculate the relative concentrations of stereosequences present in the
polymers after various extents of epimerization(17). By correla-
ting the resonance patterns of epimerized polystyrenes that were
prepared by Shepherd's procedure with stereosequence distributions
calculated for the polymers using the Monte Carlo method, assignments have been developed for the methine proton(17), methine car-
bon(18), methylene carbon(18) and aromatic C-1 carbon resonances
of polystyrene(19). These assignments agree very well with assignments developed in the model compound studies, and they lead to
the conclusion that polystyrene, as prepared by free radical polymerization methods, has a Bernoullian stereosequence distribution
with a meso content, P(m) of about 0.46. In other words, the
polymer is almost atactic.

Our previous reports concerned 20 MHz ^{13}C-NMR spectra of partially epimerized polystyrenes. The aromatic C-1 carbon resonances were recorded at room temperature and were poorly defined
when the extent of epimerization was high. A need for remeasuring
these resonances at higher temperature, using higher field spectrometers was clearly evident. This paper is therefore concerned
with the 75 MHz ^{13}C-NMR specrra of partially epimerized isotactic
polystyrenes.

Experimental

Isotactic Polystyrene. Isotactic polystyrene was prepared by
heating a 10% solution of styrene in benzene at 60° for 24 hours
in the presence of a catalyst prepared in the presence of monomer
from equimolar amounts of $(C_2H_5)_3Al$ (heptane solution) and T_1Cl_3-AA
(heptane suspension). The reaction mixture was diluted with benzene and poured into an excess of isopropanol. The resulting pre-
cipitate was dissolved in warm methylene chloride. The solution
was filtered and added to hot methyl ethyl ketone. The solution
was concentrated and cooled to obtain the isotactic polymer in the
form of a white powder that was washed with methyl ethyl ketone
and dried under vacuum. Final purification was achieved by repre-
cipitation of the polymer from benzene solution into methanol,
followed by drying at 60° under vacuum.
Epimerization Procedure. Potassium t-butoxide that had been heated
under vacuum at 130°C for three hours and hexamethylphosphoramide
that had been dried over CaH_2 at 60° for 24 hours and then distilled from $LiAlH_4$ under reduced pressure were used. Solutions containing 10g. isotactic polystyrene and 56g. (0.5 mole) KOtBu per
liter of hexamethylphosphoramide (Caution - suspected carcinogen)
were prepared under nitrogen and were heated at 100°C for periods
that ranged from 1 to 20 hours. The reaction mixtures were initially pink and ultimately developed a violet color that intensi-
fied with reaction time. The reaction mixtures were poured into
methanol to isolate the polymers. These were reprecipitated from
benzene solution into methanol and were dried under vacuum at 40°C.
The use of temperatures higher than 100°C causes extensive molecular weight reduction. The samples employed in this study are the
same ones used in our previous ^{13}C-NMR studies.
NMR Measurements. In all cases, ten percent solutions of the polystyrene samples dissolved in a 9:1 1,2,4-trichlorobenzene:nitro-
benzene-d_5 mixture were used for the NMR studies. 75 MHz ^{13}C-NMR
^1H-decoupled spectra of the epimerized polystyrene samples were
recorded at 150-160° using a Bruker WH-300 NMR Spectrometer. A
70° pulse width, an acquisition time of 0.82 seconds with a 16K
data size, and a pulse delay of 0.1 second were employed. The
number of transients collected varied from 3000 to 10,000 and the
data were processed with a line broadening of 0.8-1.0 Hz. A T_1
study done on the aromatic C-1 and aliphatic carbon resonances of
polystyrene at 200°C, using a Varian XL-400 NMR Spectrometer, re-
vealed that within experimental error the individual components of
these resonance patterns had the same relaxation time(35). This
indicates that the conditions described above are appropriate for
obtaining resonance patterns that could be analyzed quantitatively.
    Resonance area measurements were made by cutting and weighing
spectra tracings. Spectra simulations were done using a Calcomp
plotter with the aid of a program written by Dr. B.L. Bruner of
the University of Kentucky, using stereosequence distributions calculated for the polymers by Monte Carlo simulations of the epimeri-

zation process and using chemical shifts obtained by methods to be described later in this paper. A line shape that was a combination of Lorentizian and Gaussian functions was assumed for each line. This combined function has worked well in our previous simulation studies. A constant line-width was used for all lines within a given spectrum.

Monte Carlo Simulations. The epimerization reaction was simulated using the Monte Carlo program we described earlier(17). A 5000 element array was allocated to store information about the configurations of monomer units at various positions in a 5000 unit polymer chain. The positions were indexed in such a way that the polymer could be considered cyclic. This was done to avoid end group effects. The configurations (R or S) at individual sites were indicated by 0 or 1 values. The polymer chain was made isotactic by giving all elements of the array initial values of 0. The chain was then "epimerized" by conducting a series of "epimerization events," each of which involved the following steps:

1. A site on the chain was selected to experience an epimerization event by calling a random number that ranged from 1 to 5000.
2. The structure of the site selected, including structures of nearest neighbors was then determined. The possible structures were 000, 001, 101, 110, 011 and 111.
3. Depending on the structure found, a probability (P) that the site selected would have configuration 1 at the completion of the epimerization event was established by use of Table I. The V value in this table (R in previous papers) has a range of 0 - 1 and is the only parameter required for the simulation.
4. A second random number having a range of 0 - 1 was then called and compared to P. When this number was less than or equal to P, the array element representing the site was given a value of 1. Otherwise it was given a value of 0.
5. An epimerization event counter was then incremented and compared to an output schedule. If output was desired after this event, the chain (array) was analyzed for all possible dyad, triad, tetrad, pentad, hexad and heptad stereosequences and the results were stored.

Epimerization events were repeated until the chain reached equilibrium. The calculation was repeated ten times and the stereosequence concentrations evaluated after specified numbers of epimerization events were then averaged, printed and punched for use in spectra simulations.

In relating Monte Carlo and experimental results, the value of the epimerization event counter can be used as a quantity that is

Table 1: Probabilities That the Selected Site Has Configuration 1
          at The End of an Epimerization Event

Structure Found	Probability (P)
000	V
001 or 100	0.5
101	1-V
111	1-V
110 or 011	0.5
010	V

proportional to time in actual epimerization experiments. In the
present study, however, Monte Carlo results were correlated with
experimental results by matching mm-contents calculated by the
Monte Carlo method to mm-contents evaluated from the 300 MHz methine proton resonances of the polymers.

Two different V values were used for the Monte Carlo calculations. A value of 0.5 was used to predict the effect of a completely random epimerization process that yields a perfectly atactic polymer when equilibration is complete. Our previous work
employed a V value of 0.65 and this was also used here. The 0.65
value, which yields a slightly syndiotactic polymer ($P(m)=0.43$),
was selected initially because it predicts a rr-content for the
completely equilibrated polymer that agrees with theoretical pre-
dictions of Flory and Williams([36]).

Results and Discussion

Aromatic C-1 Resonance. Figures 1-6 show aromatic C-1 carbon resonance patterns observed for various epimerized isotactic polystyrene samples. The patterns are easily divided into six resonance
areas, that are designated A-F in order of increasing field. Areas
A, C and E are clear in these Figures while areas B, D and F are
cross-hatched. Except for the resonance pattern observed for atactic polystyrene (Figure 6), the resonances appear to be assignable
as follows:

    A - mmmm
    B - (mmmr + rmmm)
    C - rmmr
    D - (mmrr + rrmm) + (mmrm + mrmm)
    E - (rmrm + mrmr) + (rmrr + rrmr)
    F - rr

Thus, triad sequence distributions can be measured by combining
these fractional resonance areas: mm=A+B+C; (mr+rm)=D+E; rr=F. In
the case of the atactic polystyrene pattern (Figure 6) the demarcation between the D and E areas is not clear and it is easier to
distinguish D' and E' areas that may be assigned as follows:

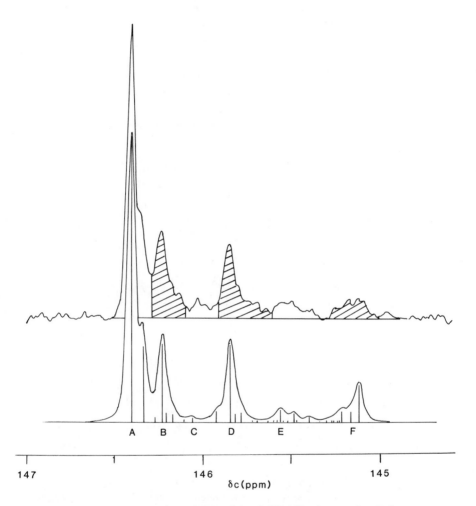

Figure 1. Observed and Simulated 75 MHz Aromatic C-1 Carbon Resonance Pattern of an Epimerized Isotactic Polystyrene Sample Having a mm-Content of 0.70.

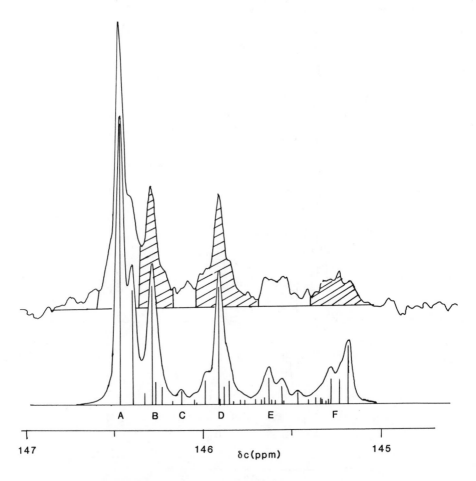

Figure 2. Observed and Simulated 75 MHz Aromatic C-1 Carbon Resonance Pattern of an Epimerized Isotactic
Polystyrene Sample Having a mm-Content of 0.59.

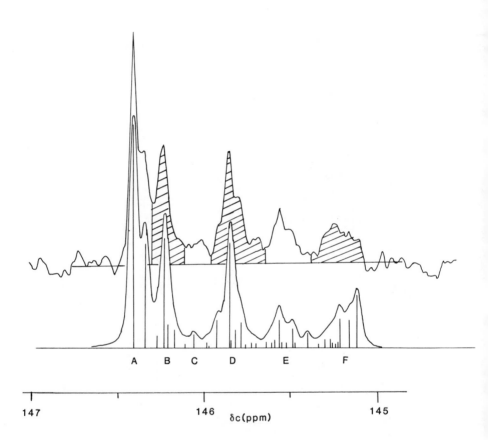

Figure 3.   Observed and Simulated 75 MHz Aromatic C-1 Carbon Resonance Pattern of an Epimerized Isotactic
Polystyrene Sample Having a mm-Content of 0.55.

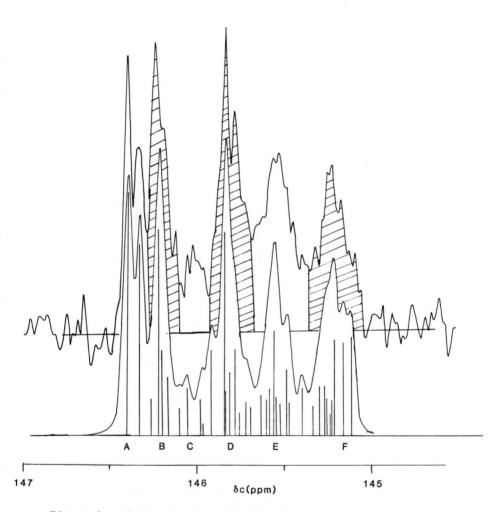

Figure 4.   Observed and Simulated 75 MHz Aromatic C-1 Car-
            bon Resonance Pattern of an Epimerized Isotactic
            Polystyrene Sample Having a mm-Content of 0.41.

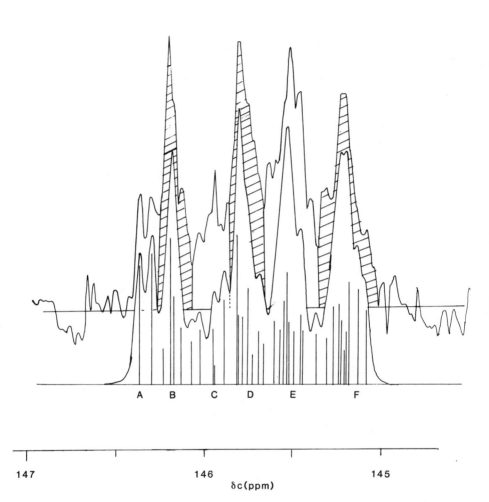

Figure 5. Observed and Simulated 75 MHz Aromatic C-1 Carbon Resonance Pattern of an Epimerized Isotactic Polystyrene Sample Having a mm-Content of 0.34.

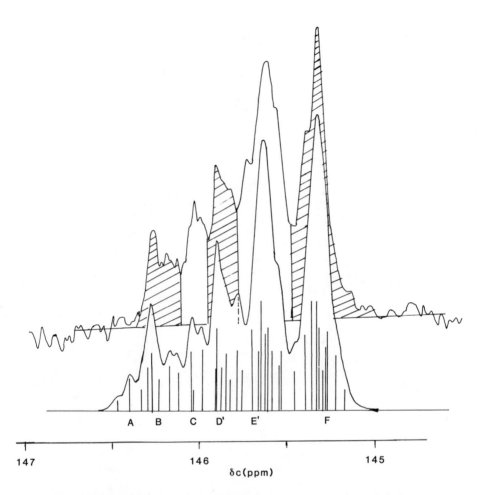

Figure 6.   Observed and Simulated 75 MHz Aromatic C-1 Carbon Resonance Pattern of Polystyrene (mm ∿0.20).

$$D' - (mmrr + rrmm)$$
$$E' - (mmrm + mrmm) + (rmrm + mrmr)$$
$$+ (rmrr + rrmr).$$

In evaluating triad stereosequence distributions for atactic polystyrene, then, the following combinations of fractional resonance
areas can be used: mm=A+B+C, (mr+rm)=D'+E'; rr=F. Table II compares triad stereosequence distributions evaluated as described
above with values calculated for the polymers by Monte Carlo simulation of the epimerization process. Calculated results based on
V values of 0.50 and 0.65 are provided for comparison. There is
little difference between calculated results based on V=0.50 or
0.65 when mm>0.35, but when mm<0.35 better correlations between
observed and calculated resonance areas (or resonance patterns,
vide infra) are obtained using V=0.65. Table III compares the
relative resonance areas of signals A-F with relative areas expected based on the pentad resonance assignments discussed above and
on stereosequence distributions calculated for the polymers using
V=0.65. The good agreement obtained between observed and calculated resonance areas indicates that the general interpretation presented above is probably correct. It should be noted that the
concentrations of mmrm and rmrm stereosequences are approximately
the same in all the polymers, so that these stereosequences could
be interchanged in the assignments proposed above without affecting
the agreement between calculated and observed resonance areas.

Table II.   Triad Stereosequence Distributions Measured for
            Epimerized Isotactic Polystyrene Samples from Aromatic
            C-1 Carbon Resonances and Calculated by Monte Carlo
            Simulation

Triad	Method	Triad Stereosequence Distributions					
mm	^{1}H	0.70	0.59	0.55	0.41	0.35	0.23
	^{13}C	0.70	0.59	0.52	0.39	0.34	0.20
	Calc^a	0.70	0.59	0.55	0.41	0.35	0.25
	Calc^b	0.70	0.59	0.55	0.41	0.35	0.19
mr+rm	^{13}C	0.24	0.29	0.34	0.44	0.45	0.50
	Calc^a	0.20	0.27	0.30	0.39	0.43	0.50
	Calc^b	0.20	0.27	0.29	0.37	0.41	0.49
rr	^{13}C	0.07	0.11	0.14	0.17	0.21	0.30
	Calc^a	0.10	0.14	0.15	0.19	0.22	0.25
	Calc^b	0.10	0.14	0.16	0.22	0.24	0.33

a) Monte Carlo Method, V=0.50
b) Monte Carlo Method, V=0.65

It is interesting that the aromatic C-1 carbon resonance patterns of the epimerized polystyrene samples are typical of the

Table III. Correlation of A–F Aromatic C-1 Carbon Resonance Areas
            Measured for Polystyrene and Epimerized Isotactic
            Polystyrene Samples with Stereosequence Distributions
            Calculated by Monte Carlo Simulation (V=0.65)

mm-Content (^{1}H-NMR)	0.70	0.59	0.55	0.41	0.34	0.20(PS)
Figure No.	1	2	3	4	5	6
Area A	0.49	0.39	0.31	0.16	0.08	–
mmmm	0.55	0.41	0.36	0.22	0.16	0.03
Area B	0.17	0.17	0.17	0.17	0.16	0.11
(mmmr+rmmm)	0.14	0.16	0.17	0.16	0.15	0.09
Area C	0.04	0.03	0.05	0.06	0.10	0.09
rmmr	0.01	0.02	0.02	0.04	0.04	0.06
Area D	0.17	0.20	0.22	0.25	0.23	0.15
mmrr+rrmm⎫ mmrm+mrmm⎭	0.16	0.20	0.21	0.23	0.24	0.12
Area E	0.07	0.09	0.11	0.19	0.23	0.35
mmrm+mrmm⎫	–	–	–	–	–	0.37
rmrm+mrmr⎬ rmrr+rrmr⎭	0.04	0.07	0.08	0.14	0.17	–
Area F	0.07	0.11	0.14	0.17	0.21	0.30
rr	0.10	0.14	0.16	0.21	0.24	0.33

patterns that are encountered in studies on the NMR spectra of
many polymers and alternating copolymers. It is not unusual to
note that the rr-centered pentad resonances occur close together,
that the mr-centered pentad resonances are spread further apart,
often occurring in two groups and that the mm-centered pentad resonances are well separated from each other. This is reasonable
if shielding by next nearest neighbor monomer units is more variable when the nearest neighbor has the same configuration as the
monomer unit under study than when the nearest neighbor has the
opposite configuration. This can be explained more clearly by
making reference to Figure 7. Suppose that shielding by nearest′
neighbor interactions causes the mm, (mr+rm) and rr resonance patterns to be well separated and let the mmmm, (mmrm+mrmm) and mrrm
resonances have chemical shifts indicated by the lines depicted at
the top of Figure 7. Now let the shielding due to next nearest
neighbor interactions be such that replacement of a terminal mm-
(or -mm) sequence by a rm- (or -mr) sequence causes a large upfield chemical shift change and let replacement of a terminal mr-
(or -rm) sequence by a rr- (or -rr) sequence cause a small down-
field chemical shift change. This would cause the complete pentad
resonance pattern to be that shown at the bottom of Figure 7,
which is analogous to that seen in Figure 5. A similar pattern
would result with other shielding tendencies; the important consi-

deration is that shielding associated with -mr (or rm-) terminal configurations must be different in magnitude (not necessarily direction) from that associated with a -rr (or rr-) terminal configuration.

Although the aromatic C-1 carbon resonance patterns observed for partially epimerized polystyrenes are readily interpreted using the above considerations, this is not the case for the pattern observed for polystyrene (or for the completely equilibrated polymer). The six-area pattern that is so clearly evident in the spectra of the partially epimerized polymers is not evident in the spectrum of polystyrene. It seems that resonances of heptads or nonads with high r-contents have chemical shifts that correspond to valleys observed in the spectra of the partially epimerized polymers. This causes the demarcation between pentad resonance patterns to become obscure when the r-content is about 0.5. This complication should disappear as the r-content increases from 0.5 and resonances due to heptads or nonads with high m-contents diminish in concentration. Unfortunately the unavailability of polystyrenes with high r-contents at the present time prevents this possibility from being pursued.

Heptad C-1 Resonances. The aromatic C-1 carbon resonance patterns contain components due to heptad stereosequences. Those evident in the spectra of polystyrenes with low r-contents are easily assigned from their relative intensities. An attempt to assign the other heptad resonances can be made by generating empirical shift rules, using procedures that have been described previously(18,19). This approach assumes that shielding of a quaternary aromatic carbon by monomer units in one direction of the polymer chain is independent of the configurations of the monomer units going in the opposite direction. Although this is probably not strictly true, it allows one to develop an initial set of assignments with a minimum number of adjustable parameters. The assignments can then be refined to obtain improved correspondence between observed and calculated resonance patterns. Seven parameters are needed to calculate the chemical shifts of 36 heptads by this approach and four of these can be evaluated directly from the resonance patterns of polymers epimerized to low extents. The seven parameters will be represented by the symbol Δxyz, where x, y and z can be r or m. Using $\delta abcdef$ to represent the chemical shift of a given heptad, where a,b,c,d,e and f can also be either r or m, the following definition prevails:

$$\Delta xyz = \delta mmmxyz - \delta mmmmmm = \delta abcxyz - \delta abcmmm, \text{ etc.}$$
$$= \delta zyxmmm - \delta mmmmmm = \delta zyxabc - \delta mmmabc, \text{ etc.}$$

$\delta mmmmmm$ can be measured from the chemical shift of isotactic polystyrene or from that of the largest signal in polymers epimerized to low extents and is 146.40 ppm for the conditions employed in the present study. Accordingly, the chemical shift of any heptad can be calculated by the following general formulas and specific examples.

$$\delta abcdef = 146.40 + \Delta cba + \Delta def$$
$$\delta mmmrrm = 146.40 + \Delta rrm$$
$$\delta rrmmmr = 146.40 + \Delta mrr + \Delta mmr$$
$$\delta mmrrmm = 146.40 + 2\Delta rmm$$

When the chemical shifts of obviously assignable resonances are known, they can be used, with equations like those written above, to evaluate Δxyz parameters. The nature of the epimerization process makes it possible to evaluate Δmmr, Δmrr, Δrrm and Δrmm directly from the spectra of polymers epimerized to low extents. Figure 8 shows a section of an isotactic polymer chain that contains one inverted monomer unit. It can be seen that (mmmmmr + rmmmmm), (mmmmrr + rrmmmm), (mmmrrm + mrrmmm) and mmrrmm heptads include the inverted monomer unit and that no other heptad stereosequences are present. Thus, resonances due to these heptads should be prominent in the spectra of isotactic polystyrenes that have been epimerized to only low extents (e.g., Figure 1 and 2). The signal due to mmrrmm heptads should be easily recognized because it must have an intensity that is one-half that of the other heptads. Provided that the other resonances occur between the mmmmmm (largest) and mmrrmm signals, it is reasonable to expect the other major resonances to occur in the following order: mmmmmm, (mmmmmr + rmmmmm), (mmmmrr + rrmmmm), (mmmrrm + mrrmmm), mmrrmm. Based on this expectation, resonances of the heptads mentioned can be identified and their chemical shifts can be used to evaluate Δmmr, Δmrr, Δrrm and Δrmm.

Evaluation of the remaining three parameters, Δmrm, Δrmr and Δrrr, is not as easily done as is the evaluation of the four parameters discussed above, but a systematic approach is possible. It is based on the observation that ten heptads, having concentrations less than those discussed above, but having very similar concentrations, make minor contributions to the spectra of polymers epimerized to low extents. These heptads are listed in Table IV. The chemical shifts of four of these heptads (lines 2,3,6 and 8) can be calculated using various combinations of Δmmr, Δmrr, Δrmm and Δrmm. The six remaining heptads fall into two groups. The chemical shifts of the members of group A (lines 1,4 and 7) will be separated from that of the mmmmmm signal by Δmrm, Δrmr and Δrrr and those of Group B (lines 5,9 and 10) will be separated from that of the (mmmrmm + mmrmmm) signal (δ_c must be calculated) by Δmrm, Δrmr and Δrrr also. These parameters can be estimated by attempting simulations of the experimental resonance pattern using calculated heptad intensities, measured chemical shifts for major heptad resonances, chemical shifts calculated for four of the minor heptad resonances, (lines 2,3,6 and 8) and variable chemical shifts for the other six resonances with the constraint that within Groups A and B the relative spacing of the resonances, must be the same. Since the intensities of the six lines in Group A and B will be approximately the same in polymers epimerized to low extents, an

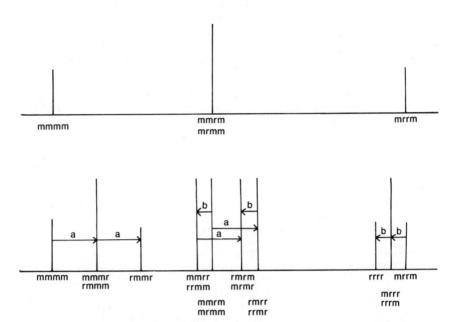

Figure 7. Influence of Long Range Shielding by Stereosequences on Resonance Patterns.

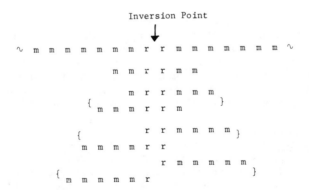

Figure 8. Heptads Present in the Vicinity of an Inverted
Monomer Unit in an Isotactic Polymer.

Table IV.   Heptad Stereosequences Responsible for Minor Signals
             in Partially Epimerized Isotactic Polystyrene Samples
             having mm-Contents Below 0.50.

Line	Heptad	Parameters Required to Calculate δ_c	Comment
1	mrmmmm + mmmmrm	Δmrm	Group A
2	rrmmmr + rmmmrr	Δmmr + Δmrr	Calculated
3	rrmmrr	2Δmrr	Calculated
4	mmmrmr + rmrmmm	Δrmr	Group A
5	mrmrmm + mmrmrm	Δmrm + Δrmm	Group B
6	rmmrrm + mrrmmr	Δmmr + Δrrm	Calculated
7	mmmrrr + rrrmmm	Δrrr	Group A
8	rrmrrm + mrrmrr	Δrrm + Δmrr	Calculated
9	rmrrmm + mmrrmr	Δrmr + Δrmm	Group B
10	rrrrmm + mmrrrr	Δrrr + Δrmm	Group B

assumption will have to be made about the relative magnitudes of
Δmrm, Δrmr and Δrrr. In the present work it was assumed that Δrmr
> Δrrr > Δmrm. Figure 9 may help to illustrate this approach.
        Once all seven Δxyz parameters have been evaluated, chemical
shifts for the remaining 21 heptads can be calculated and used for
simulating the spectra of polymers epimerized to higher extents,
as well as the spectrum of polystyrene. Minor adjustments can
then be made in calculated chemical shifts to fill in or deepen
valleys or to eliminate distortions caused by coincidences of calculated chemical shifts. This approach worked reasonably successfully when applied to quaternary aromatic carbon resonances observed for epimerized isotactic polystyrenes as measured at 20 MHz
and room temperature(19). It proved necessary, however, to shift
heptads containing a central rmrr (or rrmr) pentad upfield by
0.256 ppm to obtain good agreement between observed and simulated
spectra. This suggests that the shielding experienced by a nucleus
from one direction of the polymer chain may not be entirely independent of the structure of the chain that proceeds in the
opposite direction.
        This same general approach was used in the present study to
develop Δxyz values for polystyrene spectra recorded at 150°. The
values obtained, together with those developed previously for
spectra recorded at room temperature(19) are given in Table V.
The most significant difference between the two sets of values is
the dramatic increase in Δrrr with an increase in temperature.
Since Δrmm, Δrrm, Δrmr and Δrrr all have similar and large negative
values compared to the other Δxyz values at 150°, it is understand-
able that polystyrene spectra recorded at high temperature can
provide reliable measures of rr-triad concentrations.
        Minor adjustments were made in chemical shifts calculated
using Δxyz values to improve the quality of fit between simulated
and observed spectra. The heptad chemical shifts used to simulate

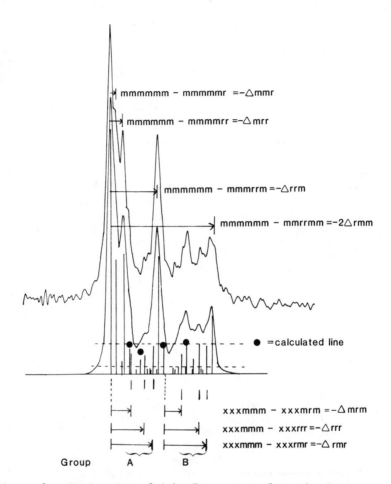

Figure 9.   Estimation of Δabc Parameters from the Spectra
of Epimerized Polymers Having High mm-Contents.

Table V. Empirical Chemical Shift Parameters (Δxyz) Evaluated
          for Polystyrene at 20° and 150°

Parameter	Value (ppm) 20°	Value (ppm) 150°
Δmmr	−0.080	−0.077
Δmrm	−0.288	−0.266
Δrmm	−0.752	−0.728
Δmrr	−0.192	−0.196
Δrrm	−0.695	−0.637
Δrmr	−0.621	−0.665
Δrrr	−0.496	−0.616
rmrr correction	−0.256	−0.112

the spectra are given in Table VI, along with those calculated
using the parameters listed in Table V. Table VI also compares
the relative order of these heptad resonance assignments with the
relative order of methyl carbon heptad resonances calculated for
polypropylene by Schilling and Tonelli(37), using the rotational
isomeric state model. The correspondence between the two sets of
assignments is surprisingly good. Although some aromatic C-1 carbon pentad resonance patterns overlap in the polystyrene spectra,
the relative ordering of the pentad resonance patterns exactly
matches that of the methyl carbon pentad resonances of polypropylene, as calculated by Schilling and Tonelli(37) [mmmm, (mmmr +
rmmm), rmmr, (mmrr + rrmm), (mmrm + mrmm), (rmrr + rrmr), (rmrm +
mrmr), rrrr, (mrrr + rrrm), mrrm, in order of increasing field].
This ordering is also in agreement with the assignments developed
for the aromatic C-1 resonances of polystyrene by Sato and Tanaka
(11). It is in only fair agreement with ordering based on
Tonelli's recent calculations for polystyrene(38), which, in contrast to the polypropylene calculations(37), tend to group the
lines into four general areas. Resonances of mm-centered pentads
are calculated to occur in three of these areas and there is more
extensive overlapping of mm- and (mr+rm)-centered pentad resonance
regions than seems reasonable based on our simulation studies.

Figures 1 - 6 compare observed aromatic C-1 carbon resonance
spectra with simulated spectra based on the heptad chemical shifts
given in Table VI and on heptad stereosequence concentrations calculated by Monte Carlo simulation of the epimerization process,
using V=0.65. The simulation spectra reproduce the general features of the observed spectra very well and can be considered to
be in at least semi-quantitative agreement with the observed spectra. The agreement between observed and simulated spectra might
be improved if spectra with higher S/N ratios were employed and if
additional parameter adjustments were made. It seems, however,
that the heptad assignments developed in this work are reasonably
correct.

Table VI. Heptad Assignments for Aromatic C-1 Carbon Resonances of Polystyrene

Heptad	δ_c(Calc)	δ_c(Simulation)	Relative Order Polystyrene[a],[b]	Polypropylene[c]
mmmmmm	146.40	146.40	1[a] 1[b]	1
rmmmmm + mmmmmr	146.32	146.32	2 2	2
rmmmmr	146.25	146.25	3 3	3
mrmmmm + mmmmrm	146.13	146.13	6 9	5
mrmmmr + rmmmrm	146.06	146.06	7 11-13	7
rrmmmm + mmmmrr	146.20	146.20	4 7,8	4
rrmmmr + rmmmrr	146.13	146.18	5 10	6
mrmmrm	145.87	145.91	10 18	10,11
rrmmrm + mrmmrr	145.94	145.92	9 17	9
rrmmrr	146.01	146.01	8 16	8
mmmrrm + mrrmmm	145.76	145.77	12 11-13	10,11
rmmrrm + mrrmmr	145.69	145.70	14 14,15	13
mmmrrr + rrrmmm	145.78	145.85	11 11-13	12
rmmrrr + rrrmmr	145.71	145.76	13 14,15	14
mmmrmm + mmrmmm	145.67	145.67	16 5	16
rmmrmm + mmrmmr	145.59	145.60	18 7,8	18
mmmrmr + rmrmmm	145.74	145.74	15 4	15
rmmrmr + rmrmmr	145.66	145.63	17 6	17
mrmrmm + mmrmrm	145.41	145.36	25 22	26
rrmrmm + mmrmrr	145.48	145.41	24 20,21	24
mrmrmr + rmrmrm	145.47	145.50	20 20,21	25
rrmrmr + rmrmrr	145.54	145.53	19 19	23
mrmrrm + mrrmrm	145.39	145.35	26 26	21
rrmrrm + mrrmrr	145.46	145.45	22 23,24	19
mrmrrr + rrrmrm	145.41	145.43	23 25	22
rrmrrr + rrrmrr	145.48	145.48	21 23,24	20
mrrrrm	145.13	145.27	27 27	27
rrrrrm + mrrrrr	145.15	145.15	29 28	28
rrrrrr	145.17	145.20	28 29	29
mrrrmm + mmrrrm	145.04	145.07	32 32	32
rrrrmm + mmrrrr	145.06	145.06	34 33	33
mrrrmr + rmrrrm	145.10	145.11	31 31	30
rrrrmr + rmrrrr	145.12	145.12	30 30	31
mmrrmm	144.94	144.94	36 36	36
rmrrmm + mmrrmr	145.01	145.00	35 35	35
rmrrmr	145.07	145.08	33 34	34

(a) Based on lines used for simulations shown in Figures 1-6.
(b) Based on calculations of Tonelli(38).
(c) Based on calculations of Tonelli and Schilling(37).

Spectra simulations based on stereosequence distributions calculated for the polymers using V=0.50 also matched the experimental spectra very well, except for polystyrene (or the completely epimerized polymer). It was not possible to develop heptad resonance assignments that afforded uniform agreement between observed and simulated spectra for all the samples studied when stereosequence distributions based on V=0.50 were used. Since resonance assignments can be made using epimerized polymers with high mm contents using stereosequence concentrations based on either V=0.50 or V=0.65, these assignments could be used to determine what V value was appropriate for polystyrene (or the completely epimerized polymer). V values ranging from 0.62 to 0.65 afforded good agreement between simulated and observed spectra of polystyrene. Assuming Bernoullian statistics, $V=(1-\sigma)^2/\sigma^2$, where σ is the probability of a meso placement in the completely equilibrated polymer (or in polystyrene). V values of 0.62-0.65 thus imply that polystyrene can be characterized by a σ value of 0.44.

One disturbing aspect of this portion of the study is the fact that the mmrrmm signal (δ_c=∿145.3 ppm) is not as intense in some observed spectra (Figures 1-4) as it should be based on simulation and on arguments presented earlier in this paper. This may indicate an influence of nonads, or a diminished sensitivity of carbons in this environment due to relaxation time or NOE differences. Additional study of this point is merited.

Methylene and Methine Carbon Resonances. Figure 10 compares the methylene and methine carbon resonance patterns observed for polystyrene and the epimerized polymers. The methylene carbon spectra are too noisy to justify quantitative study. An analysis of spectra recorded using a larger number of FID accumulations will be reported subsequently. However, the patterns observed are qualitatively similar to those discussed earlier that were recorded with a 20 MHz spectrometer(18). The methine carbon resonance patterns occurred over a small chemical shift range and were therefore adequately defined for quantitative study. We reported previously that the methine carbon resonance of these polymers occurs in two general areas and assigned the lower field area to mm-triads. Figure 11 compares the proportion of methine carbon resonance observed in this lower area with mm-contents measured for the polymers from their methine proton resonance patterns. It can be seen that there is a 1:1 correspondence between these two quantities, thus proving that the lower field methine carbon resonance is due to mm-triads.

Conclusions

Although the aromatic C-1 carbon resonance of polystyrene is very complex, relatively simple aromatic C-1 carbon resonance spectra are observed for partially epimerized isotactic polystyrenes. Studies on such "model polystyrenes" provide the information needed to interpret the spectrum of polystyrene itself. Based on assign-

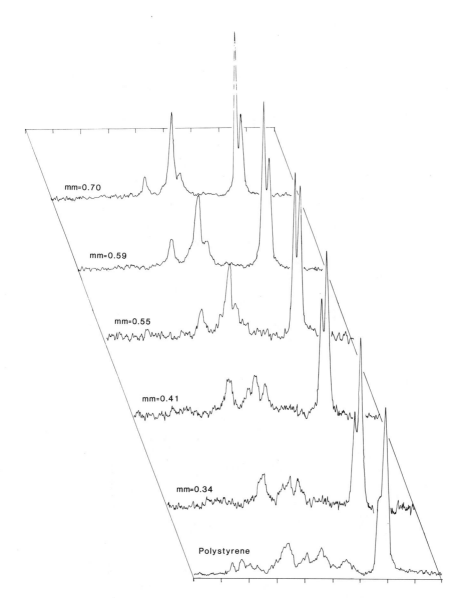

Figure 10. 75 MHz Methylene and Methine Carbon Resonance of Polystyrene and of Epimerized Isotactic Polystyrene Samples.

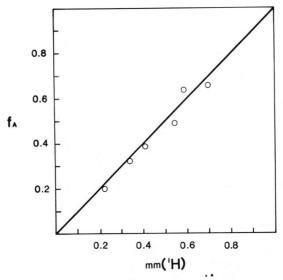

Figure 11. Correlation of Lower Field Methine Carbon
            Resonance Area (f_A) with mm–Contents of Poly-
            styrene and Epimerized Isotactic Polystyrenes.

ments made in this paper and in others in this series, the ^{1}H- and ^{13}C-NMR spectra of polystyrene indicate that it can be characterized by a σ value of ~ 0.45.

Acknowledgments

This study was supported in part by a grant from the National Science Foundation (DMA-80-10709).

Literature Cited

1. Bovey, F.A.; Tiers, G.V.D.; Filipovich, G., J. Polymer Sci. 1959 38, 73.
2. Kawamura, T.; Uryu, T.; Matsuzaki, K., Makromol. Chem., Rapid Comm. 1982, 3, 651 and references cited therein.
3. Suparno, S.; Lacoste, J.; Raynal, S.; Sledz, J.; Schue, F., Polymer J. 1981, 13, 313.
4. Jasse, B.; Laupretre, F.; Monnerie, L., Makromol. Chem. 1977, 178, 1987.
5. Nguyen-Tran, T.M.; Laupretre, F.; Jasse, B., Makromol. Chem. 1980, 181, 125.
6. Elgert, K.F.; Henschel, R.; Schorn, H.; Kosfeld, R., Polymer Bulletin 1981, 4, 105.
7. Tanaka, Y.; Sato, H.; Saito, K.; Miyashita, K., Makromol. Chem., Rapid Comm. 1980, 1, 551.
8. Sato, H.; Tanaka, Y.; Hatada, K., Makromol. Chem., Rapid Comm. 1982, 3, 175.
9. Sato, H.; Tanaka, Y.; Hatada, K., Makromol. Chem., Rapid Comm. 1982, 3, 181.
10. Tanaka, Y.; Sato, H.; Saito, K.; Miyashita, M., Rubber Chem. and Technol. 1981, 54, 686.
11. Sato, H.; Tanaka, Y., paper published in the present volume.
12. Tonelli, A.E., Macromolecules 1979, 12, 252.
13. Yoon, D.Y.; Flory, P.J., Macromolecules, 1977, 13, 562.
14. Fujiwara, Y.; Flory, P.J., Macromolecules 1970, 3, 43.
15. Trumbo, D.L.; Chen, T.K.; Harwood, H.J., Macromolecules 1981, 14, 1138.
16. Trumbo, D.L.; Suzuki, T.; Harwood, H.J., Polymer Bulletin 1981, 4, 677.
17. Shepherd, L.; Chen, T.K.; Harwood, H.J., Polymer Bulletin 1979, 1, 445.
18. Chen, T.K.; Gerkin, T.A.; Harwood, H.J., Polymer Bulletin 1980, 2, 37.
19. Chen, T.K.; Harwood, H.J., Makromol. Chem., Rapid Comm. 1983, 4, 463.
20. Randall, J.C. "Polymer Sequence Determination - Carbon-13 NMR Method," Academic Press: New York, 1977, pp 87-92, 116-119, and references cited therein.
21. Suparno, S.; Lacoste, J.; Raynal, S.; Regnier, J.F.; Schue, F.; Sempere, R.; Sledz, J., Polymer J. 1980, 12, 861.

22. Inone, Y.; Nishioka, A.; Chujo, R., <u>Makromol. Chem.</u> 1972, 156, 207.
23. Ueno, A.; Schuerch, C., <u>J. Polym. Sci., Part B</u> 1965, <u>3</u>, 53.
24. Clark, H.G., <u>J. Polym. Sci., Part C</u> 1968, <u>16</u>, 3455.
25. Flory, P.J.; Williams, A.D., <u>J. Am. Chem. Soc.</u> 1968, <u>91</u>, 3118.
26. Mercier, J.; Smets, G., <u>J. Polym. Sci., Part A</u> 1963, <u>1</u>, 1491.
27. Hogen-Esch, T.E.; Tien, C.F., <u>J. Polym. Sci., Part B</u> 1979, <u>17</u>, 431.
28. Hogen-Esch, T.E.; Tien, C.F., <u>Macromolecules</u> 1980, <u>13</u>, 207.
29. Suter, U.W.; Pucci, S.; Pino, P., <u>J. Am. Chem. Soc.</u> 1975, <u>97</u>, 1018.
30. Stehling, F.; Knox, J.R., <u>Macromolecules</u> 1975, <u>8</u>, 595.
31. Suter, U.W.; Neuenschwander, P., <u>Macromolecules</u> 1981, <u>14</u>, 528.
32. Dworak, A.; Harwood, H.J., to be published.
33. Williams, A.D.; Brauman, J.I.; Nelson, N.J.; Flory, P.J., <u>J. Am. Chem. Soc.</u> 1967, <u>89</u>, 4807.
34. Shepherd, L., Ph.D. Thesis, University of Akron, Akron, Ohio, 1979.
35. Gray, G., private communication of spectra.
36. Williams, A.D.; Flory, P.J., <u>J. Am. Chem. Soc.</u> 1969, <u>91</u>, 3111.
37. Schilling, F.C.; Tonelli, A.E., <u>Macromolecules</u> 1980, <u>13</u> 270.
38. Tonelli, A.E., <u>Macromolecules</u> 1983, <u>16</u> 604.

RECEIVED November 3, 1983

Stereospecific Polymerization of α-Olefins: End Groups and Reaction Mechanism

A. ZAMBELLI[1] and P. AMMENDOLA[1]

Universita di Napoli, Naples, Italy

M. C. SACCHI[2] and P. LOCATELLI[2]

Istituto di Chimica delle Macromolecole, Consiglio Nazionale delle Ricerche, Rome, Italy

The achievements concerning reaction mechanisms of α-olefin polymerizations are
summarized. The contributions of ^{13}C NMR
in this field, particularly concerning the
stereochemical structure of ^{13}C enriched
end groups, are discussed.

Since the late 1960's high resolution NMR became an
increasingly important method for investigating the
structure of macromolecules(1). The detailed knowledge of the molecular structure so achieved had a
large impact in the field of correlations between
structure and properties of synthetic polymers and
greatly helped the understanding of the mechanism
of polymerization reactions. Proton and ^{13}C NMR analyses have been extensively used for studying the
mechanism of stereospecific polymerization of α-ole-
fins(1). Isotopic substitution has also been very
helpful in order to simplify the spectra, to increase
the sensitivity and, even more important, in removing structure degeneracy.
    In this chapter, we will briefly summarize some
of the results reported in the literature concerning
the mechanisms of stereospecific polymerizations of
α-olefins and discuss some of the latest results
obtained by our research groups via ^{13}C NMR analyses.

[1]Current address: Via Mezzocannone 4–80134, Naples, Italy.
[2]Current address: Via Bassini 15A–20133, Milan, Italy.

Mechanism of Addition to the Double Bond

The structures of the monomer units which could be obtained by either cis or trans addition of the reactive metal-carbon bond of the active site or the double bonds of the α-olefins are degenerate. Degeneracy can be removed by a proper isotopic substitution on the monomer which is to be polymerized. For instance, cis-1-d_1-propene could give polymer chain units having different structures(2) depending on the stereochemical mechanism of addition, as shown in the following scheme (Fisher projections):

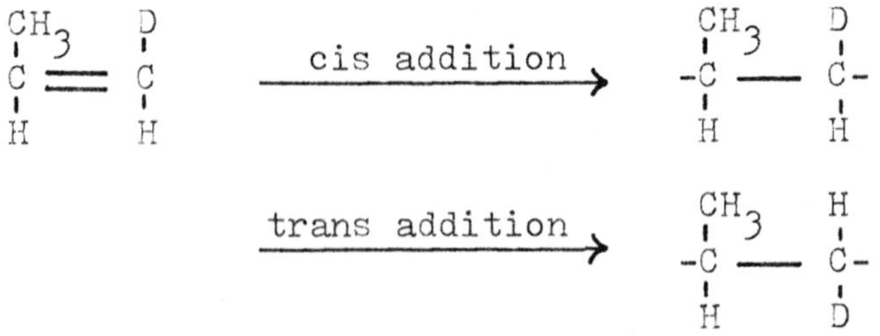

The structures of the monomer units actually resulting from both isotactic and syndiotactic Ziegler-Natta polymerizations of cis-1-d_1-propene have been determined by IR and ^1H NMR. The additions have been found to be cis(3,4).

Regiospecificity

The insertion of α-olefins on the metal-carbon bond (Mt-R) of the active sites could be either primary (or metal to C1) or secondary (or metal to C2).

$$Mt\text{-}R + C_3H_6 \xrightarrow{\text{primary insertion}} Mt\text{-}CH_2CH(CH_3)\text{-}R$$

$$Mt\text{-}R + C_3H_6 \xrightarrow{\text{secondary insertion}} Mt\text{-}CH(CH_3)CH_2\text{-}R$$

As reported in the literature, the problem of determining the actual type of insertion for syndiotactic polypropylene has been faced by observing the amount of irregularly arranged monomer units in polypropylene and in ethylene-propylene copolymers(5-7). The problem of the regiospecificity of the insertion has been recently investigated by analyzing the structure of the end groups of both syndiotactic and isotactic polypropylene. Selectively ^{13}C enriched polymers

have been prepared in the presence of different catalytic systems and analyzed by ^{13}C NMR(8). The ^{13}C
NMR spectrum of a sample of syndiotactic poly-3-^{13}C-
propylene (sample 1) prepared in the presence of the
catalytic system, VCl$_4$-Al(CH$_3$)$_2$Cl is reported in Figure 1.1. The resonances at 20.69, 20.87, 21.54, and
21.74 ppm from HMDS(9) have been attributed to the
presence of ^{13}C enriched methyls of isobutyl end groups
while the resonance at 12.37 ppm has been attributed to
^{13}C enriched methyls of n-propyl end groups. The last
resonance is also observed in the spectrum of syndiotactic poly-3-^{13}C-propylene prepared in the presence
of the catalyst, VCl$_4$-Al(C$_2$H$_5$)$_2$Cl (sample 2, Figure
1.2) while those between 20.69 and 21.74 ppm are not.
These interpretations have been based on the consideration that most polymer chains are bonded to the metal
atom of the catalytic complexes or to aluminum atoms
(after chain transfer processes). Hydrolysis of this
bond affords CH$_3$ enriched n-propyl groups. Therefore,
the insertion of the last polymerized unit of each
macromolecule is secondary. On the other hand, the
insertion of the first monomer unit (initiation) on a
metal-methyl bond gives enriched isobutyl groups. As
a consequence the initiation is primary. The initiation on a metal-ethyl bond (sample 2) should afford a
2'-^{13}C-2-methylbutyl group. In the last case the resonances of the enriched methyls overlap with the resonances of the inner monomer units (Figure 1.2). These
results agree with the hypothesis that the insertion
of propylene in syndiotactic polymerizations is a
first order Markovian process. The probability for
primary insertion is higher than for secondary insertion when it occurs on a metal-primary carbon bond.
It should also be noted that the probability of secondary insertion becomes higher when occurring on a
metal-secondary carbon bond(7,10). The interested
reader is referred to the quoted papers for a detailed
discussion. In the spectrum of isotactic poly-3-^{13}C-
propylene prepared with the catalyst, δTiCl$_3$-Al(CH$_3$)$_3$
(sample 3, Figure 1.4) only enriched isobutyl end
groups are detected. Therefore in this case the insertion of the monomer is always primary.

Mechanism of the Steric Control

Stereoregular polymerization requires that the faces
of the prochiral monomer must have a different reactivity toward one given chiral reactive site. By using
^{13}C NMR to examine the stereochemical sequences of
the configurations of the monomer units of polypropyl-

ene and ethylene-propylene copolymers prepared in the presence of different catalytic systems, it has been possible to recognize that for isotactic polymerization the steric control arises from the chiral structure of the active sites. Instead, for syndiotactic polymerization, the steric control arises from the chiral carbon of the last unit of the growing chain end(7,11). Further insight into the mechanism of the steric control has been achieved by looking at the stereochemical placement of the enriched carbon on end groups formed by initiating isotactic polymerization in the presence of ^{13}C enriched organometallic cocatalysts (12). The methyl carbons of the isobutyl end groups of polypropylene are diastereotopic(13) and the chemical shift depends on the stereochemical placement with respect to the methyl groups of the neighboring monomer units(4,13). At 22.63 MHz, four different ^{13}C resonances are detected for these methyl groups depending on the spatial relationships with the methyl groups of the following two monomer units. These four resonances are observed in the spectrum of the atactic poly-3-^{13}C-propylene (sample 4) shown in Figure 1.3. The spectra given in Figure 2 are from two samples of isotactic polypropylene prepared, respectively, in the presence of ẟTiCl₃-Al(^{13}CH₃)₃ (sample 5) and ẟTiCl₃-Al(^{13}CH₂CH₃)₃ (sample 6). In the spectrum of sample 5 (Figure 2.5) only two resonances are detected for the enriched methyl groups arising from the primary insertion of propylene on the metal-^{13}CH₃ bond of the active sites. By comparison with model compounds(8), the resonance at 21.76 ppm has been attributed to the ^{13}CH₃'s having isotactic placements with respect to the methyl substituents of the succeeding monomer units and the resonance at 20.69 ppm is assigned to the ^{13}CH₃'s having syndiotactic placements(14).

$$
\begin{array}{ccc}
\text{C} & \text{C} & ^{13}\text{C} \\
| & | & | \\
-\text{C}-\text{C}-\text{C}-\text{C}-\text{C}-\text{C}-\text{C} & \quad -\text{C}-\text{C}-\text{C}-\text{C}-\text{C}-\text{C}-\text{C} \\
& & {}^{13}\text{C}
\end{array}
$$

Since these resonances in Figure 2.5 have the same relative intensities, the insertion of the first propylene unit is stereoirregular. The subsequent insertions (on metal-isobutyl and on metal-2,4-dimethylpentyl groups) are highly stereospecific. The resonances of the enriched methylene carbons of 2-methylbutyl groups, resulting after initiation of propylene polymerization in the presence of ẟTiCl₃-Al(^{13}CH₂CH₃)₃, are observable in the spectrum of sample 6 (Figure 2.6)

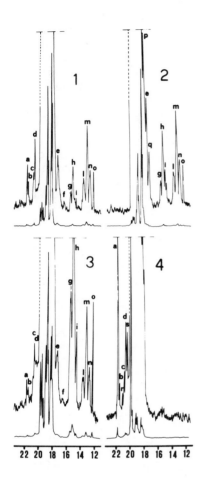

Figure 1. ^{13}C NMR spectra (22,63 MHz) of samples 1-4 (9) (methyl region). Reproduced with permission of the authors from Ref. 8.

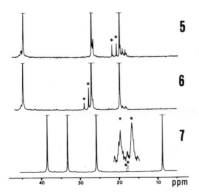

Figure 2. ^{13}C NMR spectra (22,63 MHz) of samples 5-7 (9). Reproduced with permission of the authors from Ref. 12.

at 27.7_2 ppm and 28.8_2 ppm. The resonance at 27.7_2 ppm($^{13}CH_2$ in syndiotactic placements) is twice as intense as that at 28.8_2 ppm ($^{13}CH_2$ in isotactic placements) (12). Reproduced in Figure 2.7 is the spectrum of isotactic poly-1-butene (sample 7) prepared in the presence of δ TiCl$_3$ -Al($^{13}CH_3$)$_3$. Here the 2-methyl-butyl end groups are enriched on the 2' methyl group. Two resonances at 18.1$_3$ and 17.8$_4$ ppm are detected for the $^{13}CH_3$'s in different stereochemical placements(8). They have the same intensity. As discussed in detail in references 12 and 14, these data can be explained by assuming that isotactic control of α-olefin polymerization is due to the presence of chiral active sites. However the extent of steric control increases **with** an increasing size of the alkyl group (methyl < ethyl < iso-butyl) on which the insertion of the monomer unit occurs. Work is in progress concerning the insertion of propylene between metal-phenyl bonds. We can anticipate that this insertion of propylene will be highly isotactic specific(15).

Stereoselective Polymerization of Chiral α-Olefins

As discovered by Natta, Pino and coworkers(16,17), isotactic polymerization of chiral α-olefins is stereoselective (e.g., isotactic poly-(R,S)-3-methyl-1-pentene consists of enantiomeric macromolecules which can be partially resolved.) If one considers that the isotactic steric control arises from the enantioselectivity of the chiral catalytic sites toward the enantiotopic carbon of the monomer(7,11), then stereoselectivity simply means that the insertion is diastereoselective. In other words, the diastereotopic faces of the monomer must have a different reactivity in the insertion (Figure 3). Figure 4 shows the ^{13}C NMR spectrum of isotactic poly-(R,S)-3-methyl-1-pentene obtained in the presence of δ TiCl$_3$-Al($^{13}CH_3$)$_3$ (sample 8). The resonances at 13.2$_4$, 13.5$_7$, 15.0$_9$ and 15.2$_7$ ppm are due to the $^{13}CH_3$'s of 2'-^{13}C-2,3-dimethyl-pentyl end groups formed in the initiation step(18).

$$Mt-^{13}CH_3 + CH_2=CH-CH(CH_3)-C_2H_5 \longrightarrow$$
$$Mt-CH_2-CH(^{13}CH_3)-CH(CH_3)-C_2H_5$$

According to reference 18, the steric placement of ^{13}C methyl group with respect to the 3' methyl group (see Figure 3) can be evaluated from the NMR chemical

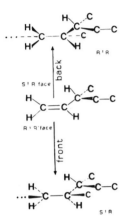

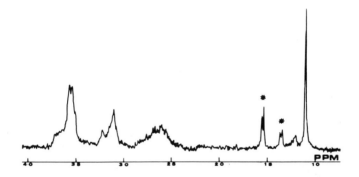

Figure 3. Diastereomeric end groups result from the attack of the diastereotopic faces of 3-methyl-1-pentene. For the configurational notation see Ref. 18. Reproduced with permission of the authors from Ref. 18.

Figure 4. ^{13}C NMR spectrum (22,63 MHz) of sample 8 (9). Reproduced with permission of the authors from Ref. 18.

shift. The resonances at 15.0_9 and 15.2_7 arise from $^{13}CH_3$'s in an erythro relationship (S'R unit on bottom of Figure 3) with respect to the 3' methyl groups and those at 13.2_4 and 13.5_7ppm correspond to the analogous threo relationship (R'R unit at the top of Figure 3). In the spectrum of Figure 4, one can observe that the intensity of the resonances of the erythro $^{13}CH_3$'s is, roughly, twice that of those of the threo $^{13}CH_3$'s. The placement of the $^{13}CH_3$'s on the end groups is related to the face of the monomer which reacted as shown in Figure 3. The conclusion reached is that the more re-active faces of the monomer are those which, after co-ordination to a transition metal atom, give a new asym-metric carbon having the same absolute configuration as that of the substituent (the front face of Figure 3).

Conclusion

The stereospecific polymerization of α-olefins is one of the best examples which illustrate the possible ap-plications of high resolution NMR to the determination of reaction mechanisms. As a matter of fact, the stereoregular structure of the reaction products and the sensitivity of ^{13}C chemical shifts to stereochemi-cal environments make it possible to obtain consider-able and important information about very subtle de-tails of the reaction mechanism. This is especially true when NMR is used in conjunction with isotopic substitution.

Literature Cited

1. Bovey, F. A. "High Resolution NMR of Macromole-cules"; Academic Press: New York, 1972.
2. Natta, G.; Farina, M.; Peraldo, M. Chim. Ind. (Milan) 1960, 42, 255.
3. Mijazawa, T.; Ideguchi, T. J. Polymer Sci. 1963, B1, 389.
4. Zambelli, A.; Giongo, M. G.; Natta, G. Makromol. Chem. 1968, 112, 183.
5. Zambelli, A.; Tosi, C.; Sacchi, M. C. Macromole-cules 1972, 5, 649.
6. Asakura, T.; Ando, I.; Nishioka, A.; Doi, Y.; Keii, T. Macromol. Chem. 1977, 178, 791.
7. Zambelli, A.; Bajo, G.; Rigamonti, E.; Makromol. Chem. 1978, 179, 1249.
8. Zambelli, A.; Locatelli, P.; Rigamonti, E. Macro-molecules 1979, 12, 156.
9. All of the chemical shifts reported in this paper are with respect to the HMDS scale. Each of the

polymers have been examined in 1,2,4-trichloro-
benzene solution at high temperature. The reader
is referred to the references reported throughout
this paper for specific experimental details.

10. Zambelli, A.; Allegra, G. Macromolecules 1980, 13,
    42.
11. Zambelli, A. in "NMR Basic Principles and Progress"
    Springer: Heidelberg, 1971; Vol. 4, p. 101.
12. Zambelli, A.; Sacchi, M. C.; Locatelli, P.;
    Zannoni, G. Macromolecules 1982, 15, 211.
13. Carman, C. J.; Tarpley, A. R. Jr.; Goldstein, J.H.
    Macromolecules 1973, 6, 719.
14. Zambelli, A.; Locatelli, P.; Bajo, G. Macromole-
    cules 1979, 12, 154.
15. Unpublished data from our laboratories.
16. Natta, G.; Pino, P.; Mazzanti, G.; Corradini, P.;
    Giannini, U. Rend. Acc. Naz. Lincei (VIII), 1955
    19, 397.
17. Pino, P.; Ciardelli, F.; Montagnoli, G. J. Poly-
    mer Sci. Part C, 1969, 16, 3256.
18. Zambelli, A.; Ammendola, P.; Sacchi, M. C.;
    Locatelli, P.; Zannoni, G. Macromolecules 1983,
    16, 341.

RECEIVED October 24, 1983

Structural Characterization of Naturally Occurring *cis*-Polyisoprenes

YASUYUKI TANAKA

Department of Material Systems Engineering, Faculty of Technology, Tokyo University of Agriculture and Technology, Koganei, Tokyo 184, Japan

The chemical structure of naturally occurring *cis* polyisoprenes was determined by 13C NMR spectroscopy using acyclic terpenes and polyprenols as model compounds. The arrangement of the isoprene units along the polymer chain was estimated to be in the order: dimethylallyl terminal unit, three *trans* units, a long block of *cis* units, and *cis* isoprenyl terminal unit. This result demonstrates that the biosynthesis of *cis*-polyisoprenes in higher plants starts from *trans*,*trans*,*trans*-geranylgeranyl pyrophosphate.

Polyisoprenes are synthesized by thousands of plant species covering most generic families. Usually, the polyisoprenes are rubbery and have predominantly a *cis*-1,4 structure. Only a relatively few plant species produce *trans*-1,4 polyisoprenes. In early studies the presence of the 3,4 structure in natural rubber was estimated at 1-2 % by infrared analysis. This conclusion was challenged after studies utilizing 1H NMR spectroscopy, which reported that natural rubber (from *Hevea brasiliensis*) and gutta percha are at least 99.0-99.5 % *cis*-1,4 and *trans*-1,4 polyisoprenes. The 3,4 isomeric structure was present in less than the minimum detectable amount (1,2). Guayule rubber from *Parthenium argentatum* was found to have a structure nearly 100 % *cis*-1,4 and identical to that of natural rubber as determined using 300 MHz 1H NMR spectroscopy (3). The differences in physical properties between natural rubber and synthetic *cis*-1,4 polyisoprenes have been ascribed in part to the structural purity of the repeating units. Synthetic polyisoprenes prepared with Al-Ti catalysts were found to be 99% *cis*-1,4, the remainder being 0-0.7 % *trans*-1,4 and 0.3-1.0% 3,4. The presence of branches and cross-linking has been considered to be another characteristic of natural rubber. However, little structural information is known about the terminal units, branching, and so-called abnormal groups in natural rubber and how they relate to specific physical properties.

0097-6156/84/0247-0233$06.00/0

The biosynthesis of natural rubber has been studied from the viewpoint of an elucidation of initiation and propagation mechanisms mainly by tracer techniques. The steps in the formation of isopentenyl pyrophosphate from acetyl–coA via mevalonate are now well established in the *in vitro* synthesis of rubber. It has also been confirmed that chain extension occurs on the surface of existing rubber particles by successive additions of isopentenyl pyrophosphate to build up chains of 5000–7000 isoprene units (4,5). The initiation step of rubber formation, however, remains unknown due to the lack of detailed information concerning the direct precursor of the chain extension.

This paper describes the structural analysis of naturally occurring *cis*-1,4 polyisoprenes using 13C NMR spectroscopy. First, the structural characterization of polyprenols, which are linear isoprenoid compounds containing 30 to 100 carbons, was carried out on the basis of information obtained from acyclic terpenes having various *cis* and *trans* isoprene units as model compounds. This method was also applied to the structural analysis of polyisoprenes. The elucidation of the structure of the end groups and the arrangement of isoprene units provides information on the mechanism of the biosynthesis of polyprenyl compounds in nature.

13C NMR Analysis of Model Compounds

By analogy with the mechanism of the biosynthesis of farnesyl pyrophosphate and all-*trans* terpenoids, it was deduced that rubber formation proceeds by the successive addition of isopentenyl pyrophosphate to dimethylallyl pyrophosphate (4).

$$\begin{aligned}&CH_3\\&\!\!\!\diagdown\\&C=CHCH_2OPP + n\ CH_2=\overset{\overset{\displaystyle CH_3}{|}}{C}CH_2CH_2OPP \longrightarrow\\&CH_3\diagup\end{aligned}$$

$$\begin{aligned}&CH_3\\&\!\!\!\diagdown\\&C=CHCH_2-(CH_2\overset{\overset{\displaystyle CH_3}{|}}{C}=CHCH_2)_n-OPP \qquad (I) \quad (OPP:\ pyrophosphate)\\&CH_3\diagup\end{aligned}$$

According to this mechanism, natural rubber chains are expected to have one dimethylallyl terminal unit and one isoprenyl pyrophosphate terminal unit; the latter may give rise to a hydroxyl group by hydrolysis. From this point of view, acyclic terpenes in the generalized structure (II) may be appropriate models for the structural characterization of natural polyisoprenes by 13C NMR spectroscopy.

$$\begin{aligned}&CH_3\\&\!\!\!\diagdown\\&C=CHCH_2-(CH_2\overset{\overset{\displaystyle CH_3}{|}}{C}=CHCH_2)_{n-2}- -CH_2\overset{\overset{\displaystyle CH_3}{|}}{C}=CHCH_2OH \qquad (II)\\&CH_3\diagup\end{aligned}$$

ω–terminal   internal α–terminal

Table I. Model compounds for 13C NMR assignment of polyisoprenoid compounds (8).

n=2	n=3	n=4
ω-T-OH	ω-T-T-OH	ω-T-T-T-OH
(geraniol)	(farnesol)	(geranylgeraniol)
ω-C-OH	ω-T-C-OH	ω-C-T-T-OH
(nerol)	ω-C-T-OH	ω-T-C-T-OH
	ω-C-C-OH	ω-C-C-T-OH
		ω-T-T-C-OH
		ω-C-T-C-OH
		ω-T-C-C-OH
		ω-C-C-C-OH

The abbreviation ω, T, and C correspond to ω-terminal, *trans*, and *cis* units, respectively.

Geraniol and nerol (n=2), farnesol isomers (n=3), and geranylgeraniol isomers (n=4) having various combinations of *trans* and *cis* isoprene units were used as model compounds, as listed in Table I. Here, the geometric isomers of farnesol were isolated from synthetic farnesol, which is a mixture of four isomers, by liquid chromatography (6). In a similar way, the geranylgeraniol isomers were separated from a mixture prepared by the isomerization of naturally occurring *trans*,*trans*,*trans*-geranylgeraniol under UV irradiation (7).

The aliphatic carbon signals, observed at 50.1 MHz in these compounds, were assigned through a consideration of chemical shift correlations among these compounds as well as by the usual 13C NMR techniques (8). In these compounds, the methyl carbon atoms in the internal *trans* and *cis* units resonated at 16.0 and 23.4 ppm, respectively, while those in the ω-terminal unit resonated at 17.6 and 25.6 ppm.

The C-1 methylene carbon atoms in the *cis* and *trans* units showed signal splittings reflecting the geometric isomerism of the unit linked to the C-1 methylene carbon atom, that is, reflecting the dyad sequences of *cis* and *trans* units. Here, the carbon atoms are designated as follows:

$$
\begin{array}{c}
5 \\
\overset{|}{C} \\
-C-C=C-C- \\
1\ 2\ 3\ 4
\end{array}
$$

The C-1 methylene carbon atoms in *trans* units resonated at 39.6 ppm (*trans*-***trans*(α)**), 39.7-39.8 ppm (ω-***trans*** and *trans*-*trans*), 39.9 ppm (*cis*-*trans*(α)), and 40.0 ppm (*cis*-*trans*), while those in *cis* units resonated at 32.0-32.1 ppm (*trans*-***cis***, ω-***cis***, and *trans*-***cis*(α)**) and 32.3-32.4 ppm (*cis*-***cis*** and *cis*-***cis*(α)**). In these correlations it was observed that the ω-terminal unit has the same shielding effect on the subsequent C-1 methylene carbon atom as the internal *trans* unit.

The C-4 methylene carbons in the *trans* and *cis* α-terminal units gave signals at 59.4 and 59.1 ppm, respectively. The geometric isomerism of the unit linked to the ω-terminal unit could be determined by the observed chemical shift of the ω C-2 carbon atom; the ω-*trans* linkage showed about 0.2-0.4 ppm upfield shift compared with that in the ω-*cis* linkage. The assignment of the signals characteristic of the terminal units and the alignment of isoprene units in the geranylgeraniol isomers are listed in Table II.

Structural Characterization of Polyprenols

Polyisoprenoid alcohols consisting of 9 to 20 isoprene units have a widespread occurrence as indicated by their presence in the leaves of higher plants, mammalian tissues, and microorganisms. Most of the polyprenols isolated from higher plants consist of *trans* and *cis* isoprene units with the exception of solanesol, which is composed of all-*trans* isoprene units. The arrangement of *trans* and *cis* units in these polyprenols has been determined from a consideration of the mechanism of the formation of betula-prenols, C(30)-C(45), isolated from wood tissue of *Betula verrucosa* (9). However, up to the present there has been no direct evidence to prove the location of the internal *trans* and *cis* units.

The arrangement of the *trans* and *cis* units in typical polyprenols was determined according to information obtained from the 13C NMR study of acyclic terpenes mentioned above. Poly-prenol-11 C(55), isolated from the leaves of *Ficus elastica*, was found to contain the ω-terminal unit, three internal *trans* units, six internal *cis* units, and a *cis* α-terminal unit from 1H NMR observations. Polyprenol-12 C(60), isolated from *Ficus elastica* and silkworm feces, showed a similar composition except that there were seven internal *cis* units.

As shown in Fig. 1, the signal due to the C-1 methylene carbon atom in *cis* units is split into two peaks at 32.37 and 32.12 ppm, which were assigned to the *cis* units in the *cis-cis* (*cis-cis*(α)) and *trans-cis* (ω-*cis*) linkages, respectively. On the other hand, the C-1 methylene carbon atom in *trans* units showed a single peak at 39.86 ppm, corresponding to the *trans* unit in the *trans-trans* (ω-*trans*) linkage. The presence of the ω-*trans* linkage was confirmed by the characteristic C-2 olefinic carbon signal of the ω-terminal unit at 131.20 ppm. The CH_2OH signal at 59.00 ppm indicated that the α-terminal unit has the *cis* configuration. The observed intensity ratio of the *trans-trans* (ω-*trans*), *trans-cis*, and *cis-cis* (*cis-cis*(α)) signals was 2.8 : 0.9 : 6.3 for polyprenol-11 and 2.7 : 1.2 : 7.1 for polyprenol-12.

Table II. Chemical shifts of 13C NMR signals in geranylgeraniol isomers (8).

Isomer*	α C-2 (=CH-)		α C-4 (-CH2OH)		trans C-1 (-CH2-)				cis C-1 (-CH2-)	
	ωC	ωT	T	C	CT	CTα	TT|ωT|	TTα	CCα / CC	TCα / TC / ωC
CCC-OH	131.55			59.08					32.30	32.04
TCC-OH		131.29		59.06			39.78		32.29	31.97
CTC-OH	131.49			59.06	40.01					32.03
TTC-OH		131.24		59.08			39.72	39.76		32.07
CCT-OH	131.51		59.47			39.89			32.34	32.03
TCT-OH		131.30	59.43			39.88	39.78			32.01
CTT-OH	131.47		59.39		40.01			39.58		32.03
TTT-OH		131.22	59.43				39.76	39.60		32.03

*The abbreviations ω, T, and C correspond to ω-terminal, *trans*, and *cis* units, respectively. The subscript α means the unit with hydroxyl group.

These signals showed T_1 value of less than 1 s and nearly full NOE values.

On the basis of these findings, it is concluded that polyprenols-11 and -12 are composed of the ω-terminal, three internal *trans*, six or seven internal *cis*, and *cis* α-terminal units aligned in that order as shown below:

$$
\begin{array}{c}
CH_3 \\
^{CH_3}C=CCH_2-(CH_2\underset{H}{\overset{H_3C}{C}}=CCH_2)_3-(CH_2\underset{}{\overset{H_3C\ H}{C}}=CCH_2)_{6,7}- -CH_2\overset{H_3C\ H}{C}=CCH_2OH
\end{array}
\qquad (III)
$$

ω-terminal   *trans*   *cis*   *cis* α-terminal

This alignment of the *trans* and *cis* units clearly indicates that these polyprenols are synthesized *in vivo* by the *cis* addition of isopentenyl pyrophosphate to *trans,trans,trans*-geranylgeranyl pyrophosphate (10).

Similarly, polyprenols-15 to -20, isolated from the leaves of *Ginkgo viloba*, showed the same splitting of C-1 methylene carbon signals, as shown in Fig. 2. From the intensity ratios of these signals it was found that these polyprenols contained two internal *trans* and 11 to 16 internal *cis* units aligned as given in (IV) (11).

$$
\begin{array}{c}
CH_3 \\
^{CH_3}C=CCH_2-(CH_2\underset{H}{\overset{H_3C}{C}}=CCH_2)_2-(CH_2\overset{H_3C\ H}{C}=CCH_2)_{11-16}- -CH_2\overset{H_3C\ H}{C}=CCH_2OH
\end{array}
\qquad (IV)
$$

ω-terminal   *trans*   *cis*   *cis* α-terminal

The polyprenols composed of *trans* and *cis* isoprene units isolated so far have been classified into two types: the ficaprenoltype and the betulaprenol-type, containing three and two internal *trans* units, respectively (12). The findings described above suggest that *trans,trans,trans*-geranylgeranyl pyrophosphate or *trans,trans*-farnesyl pyrophosphate acts as an initiator of the biosynthesis of polyprenols.

Structure of Naturally Occurring *Cis*-Polyisoprenes

The chemical shifts of the characteristic carbon signals in acyclic terpenes, polyprenols, and *cis-trans* isomerized polyisoprenes are plotted in Fig. 3. Here, the chemical shifts are correlated using the ω C-5 methyl carbon signal at 17.66 ppm as an internal standard (except for isomerized polyisoprenes) in order to compensate for the effect of solution concentration. It is clear that these chemical shifts are independent of the chain length of the compounds and can be used for the determination of the arrangement of isoprene units as well as the terminal units in various isoprenoid compounds (8).

Cis-polyisoprene isolated from the leaves of goldenrod (*Solidago altissima*) was separated into two fractions by GPC, S-1 ($\overline{Mn}$=76,000) and S-2 ($\overline{Mn}$=120,000). The 13C NMR spectra of both

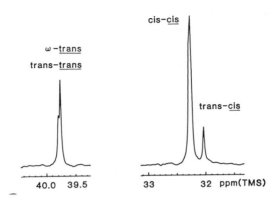

Figure 1.   C-1 methylene carbon signals in polyprenol-11 from *Ficus elastica*.   Reproduced with permission from Ref. 10. Copyright 1979, The Biochemical Society.

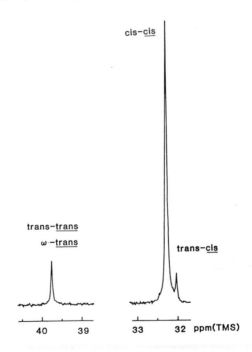

Figure 2.   C-1 metylene carbon signals in polyprenol-18 from *Ginkgo viloba*.   Reproduced with permission from Ref. 11.   Copyright 1983,   The Biochemical Society.

fractions showed five signals corresponding to the carbon atoms in
the *cis* isoprene unit.   In the expanded spectrum of S-1, obtained
by 34,730 accumulations, small signals were observed, as shown in
Fig. 4.   The signal at 59.00 ppm was assigned to the CH_2OH carbon
atom in the *cis* α-terminal unit from the relationship shown in
Fig. 3.   Similarly, the signals at 17.67 and 16.00 ppm were
ascribed to the C-5 methyl carbon atoms in the ω-terminal and
internal *trans* units, respectively.   The C-1 methyl carbon signal
due to the internal *trans* units was observed at 39.74 ppm, which
is characteristic of the *trans* units in the *trans-**trans*** and
ω-***trans*** linkage.   The C-2 olefinic carbon atom in the ω-terminal
unit resonated at 131.14 ppm, showing the presence of the ω-*trans*
linkage.   The absence of the signal characteristic of the
*cis-**trans*** linkage, which is presumed to resonate around 39.9 ppm,
indicates that all the *trans* units are situated following the
ω-terminal unit.   Although the C-1 methylene carbon signal due to
the *trans-**cis*** linkage was not observed because of overlap with the
strong signal from the *cis-**cis*** linkage, the findings shown above
strongly support the idea that a block of *cis* units is linked to
the ω-(*trans*)_m - sequence.   Similarly, small signals due to the
ω-terminal, *cis* α-terminal, and internal *trans* units were observed
in the spectrum of the sample S-2.

The relative intensities of these signals, together with the
NOE and T_1 values, are listed in Table III.   The intensity ratio
from the signals of the ω-terminal unit (ω C-5) and α-terminal
unit (α C-4) was 0.8-0.9 for both samples.   The intensity ratio of
the corresponding signals in polyprenol-11 was 0.84, in which case
the difference is ascribed to a T_1 of 7.5 versus 1.9 s for the
ω C-5 and α C-4 carbon atom, respectively, as well as to the NOE
values.   Therefore, it seems resonable to assume that polymers S-1
and S-2 contain the same amount of ω- and α-terminal units.   The
number of internal *trans* units determined from the intensity ratio
of the *trans* C-1 and α C-4 methylene carbon signals was 3.4 and
3.6 for S-1 and S-2, respectively.   On the other hand, 2.5 *trans*
units were estimated for both samples from the intensity ratio of
the *trans* C-5 and α C-4 carbon signals.   The former is considered
to be a more reliable value in view of the fact that the C-5
carbon atom shows smaller NOE and longer T_1 values than the *trans*
C-1 methylene carbon atom in polyprenols-11 and -12.   This
indicates that the polymers contain approximately three to four
internal *trans* units.

The number of the internal *cis* units determined from the
intensity ratio of either the *cis* C-1 or C-4 methylene signal to
the α C-4 methylene carbon signal, was approximately 1000 and 2200
for the polymers S-1 and S-2, respectively.   Similar NOE and
shorter T_1 values for these carbon atoms indicate that an almost
quantitative evaluation is possible for the intensity ratios
between these methylene carbon signals.   The degree of polymerization determined from the α-terminal unit agrees with that from
GPC for these samples, i.e., 1100 for S-1 and 1800 for S-2.   This

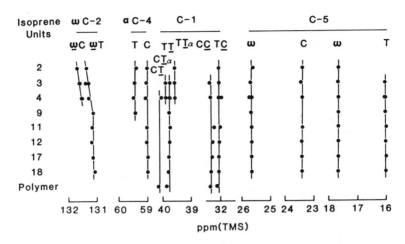

Figure 3.   Chemical shifts of the signals characteristic
of the arrangement of the isoprene units (8, 13).

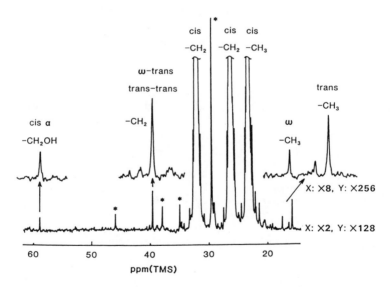

Figure 4.   Aliphatic carbon signals in *cis*-polyisoprene
from *Solidago altissima* observed with a pulse repetition
time of 7 s for 45° pulse (*denotes signals due to
impurities). Reproduced with permission from Ref. 13.
Copyright 1983, The American Chemical Society.

Table III. Relative intensities of the aliphatic carbon signals
(13). (*Values determined for polyprenol-11 as a model compound).

Chemical shift (ppm)	Assignment	Relative intensity		NOE	T_1 (s)
		S-1	S-2		
59.00	α C-4	1	1	2.62*	1.9*
39.74	*trans* C-1	3.4	3.6	2.62*	0.9*
32.23	*cis* C-1	1000	2200	2.71	0.7
26.43	*cis* C-4	1000	2200	2.78	0.8
23.41	*cis* C-5	980	2100	2.49	1.7
17.67	ω C-5	0.8	0.9	2.22*	7.5*
16.00	*trans* C-5	2.5	2.5	2.22*	3.9*

is important evidence that these polymers are linear polyisoprenes
having ω- and α-terminal units.

These facts clearly indicate that the *cis*-polyisoprene
isolated from goldenrod has the alignment of isoprene units as
shown in (V) (13).

$$
\begin{array}{cccc}
& H & H_3C & H_3C\ H & H_3C\ H \\
CH_3\diagdown & | & | & | \qquad | & | \qquad | \\
& C=CCH_2-(CH_2C=CCH_2)-(CH_2C=CCH_2)-CH_2C=CCH_2OH & & \\
CH_3\diagup & & H \qquad_m & \qquad_n &
\end{array}
\qquad (V)
$$

ω-terminal   *trans*   *cis* *cis* α-terminal

(m=3-4 and n=1000-2200)

A similar alignment of isoprene units was observed for
cis-polyisoprene isolated from the latex of *Ficus elastica*. This
is significant evidence clarifying the detailed mechanism of
rubber formation in higher plants. The presence of the sequence
consisting of three to four *trans* units linked to the ω-terminal
unit demonstrates that the primer of *cis*-polyisoprene is a prenyl
pyrophosphate possessing an all-*trans* configuration. Geranylgeraniol is one of the most common prenyl compounds in nature,
whereas geranylfarnesol (C(25) *trans,trans,trans,trans*) occurs
only rarely. Therefore, it can be presumed that the chain
extension to the *cis* polymer occurs by the addition of isopentenyl
pyro-phosphate to all-*trans* geranylgeranyl phrophosphate, as shown
in Fig. 5. It is worth noting that this structure is identical to
that of ficaprenol-type polyprenols isolated mainly from the leaf
tissues of angiosperms, except for the number of *cis* units.

On the other hand, the low molecular weight fraction
($\overline{Mn}$=10,000) of natural rubber, which was obtained by fractionation
of Hevea rubber purified by deproteinization of a commercial
latex, showed small signals characteristic of the C-1 methylene
and C-5 methyl carbon atoms in internal *trans* units at 39.80 ppm
and 16.02 ppm, respectively, as shown in Fig. 6. The signals due
to the terminal units were not detected at 59.0 ppm and 17.7 ppm,

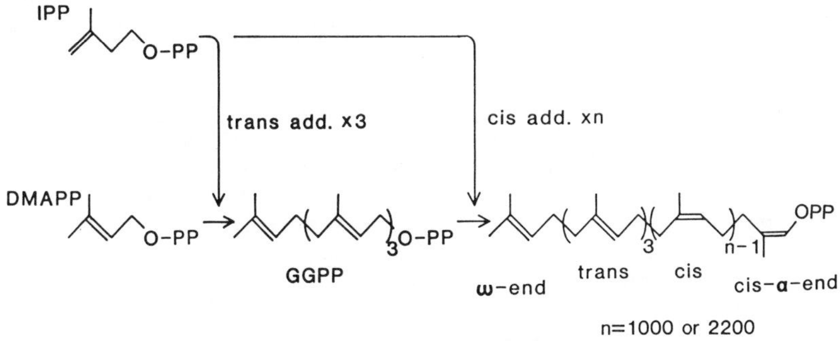

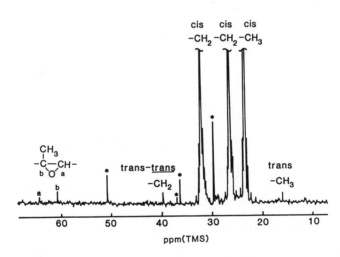

Figure 5.   Biosynthesis mechanism of *cis*-polyisoprenes (13, 14).

Figure 6.   Aliphatic carbon signals in *cis*-polyisoprene from *Hevea brasiliensis* (natural rubber) (*denotes signals due to impurities) (14).

whereas the signals from epoxide group were observed at 64.51 and 60.76 ppm. The *trans* C-1 methylene carbon signal showed the chemical shift characteristic of the *trans -trans* and ω-*trans* linkages and had a relative intensity of approximately 1/500 compared to the *cis* C-1 methylene carbon signal. Taking into account the degree of polymerization of the sample, one expects about three *trans* units per molecule to occur as an isolated sequence (14).

These facts suggest that natural rubber is synthesized *in vivo* in a similar manner to goldenrod rubber and that the terminal units subsequently undergo some particular reaction to form functional groups. One may speculate that the reactive functional groups are responsible for the formation of branches and cross-links, which are believed to occur in significant amounts in natural rubber. Very recently, Archer *et al.* found that all-*trans* geranylgeranyl pyrophosphate can act as an initiator to synthesize new rubber molecules (15). This result strongly supports the suggested mechanism for the formation of natural rubber as described above.

Literature Cited

1. Golub, M. A.; Fuqua, F. A.; Bhacca, N. S. J. Am. Chem. Soc. 1962, 84, 498 (1962).
2. Chen, H. Y. J. Polym. Sci. 1966, B4, 891.
3. Campos-Lopez, E.; Palacios, J. J. Polym. Sci. Polym. Lett. Ed. 1976, 14, 1561.
4. Lynen, F.; Henning, U. Angew Chem. 1960, 72, 820.
5. Archer, B. L.; Ayrey, G.; Cockbain, E. G.; McSweeney, G. P. Nature (London) 1961, 189, 663.
6. Sato, H; Kageyu, A.; Miyashita, K.; Tanaka, Y. J. Chromatogr. 1982, 237, 178.
7. Tanaka, Y. ; Sato, H.; Kageyu, A. Polym. Prep. Japan 1981, 30, 1834.
8. Tanaka, Y.; Sato, H.; Kageyu, A. Polymer 1982, 23, 1087.
9. Wellburn, A. R.; Hemming, F. W. Nature (London) 1966, 212, 1634.
10. Tanaka, Y.; and Takagi, M. Biochem. J. 1979, 183, 163.
11. Ibata, K.; Mizuno, M.; Takigawa, T.; Tanaka, Y. Biochem. J. 1983, 213, 305.
12. Hemming, F. W. "Biochemistry of Lipids (Biochemistry Series 1)"; Goodwin, T. W., Ed; Butterworth: London, 1974; Vol. 4, p. 39-98.
13. Tanaka, Y.; Sato, H.; Kageyu, A. Rubber Chem. Technol. 1983, 56, 299.
14. Tanaka, Y. Polym. Prep. Japan 1983, 32, 75.
15. Archer, B. L., personal communication.

RECEIVED November 3, 1983

A ^{13}C NMR Study of Radiation-Induced Structural Changes in Polyethylene

J. C. RANDALL—Phillips Petroleum Company, Bartlesville, OK 74004

F. J. ZOEPFL—Pickard, Lowe and Garrick, Inc., Washington, DC 20036

JOSEPH SILVERMAN—University of Maryland, College Park, MD 20742

Polyethylenes, irradiated to doses just short of the gel dose, have been examined for structural changes using ^{13}C NMR in a study of the effects of ionizing radiation on polyethylene. Radiation gelled polyethylenes do not produce high resolution NMR resonances possibly because NMR dipolar interactions in the gel phase are not effectively averaged by the available molecular motions. One of the major structural units formed during irradiation of polyethylene is the Y type of long chain branch. Other structural entities monitored and followed versus radiation conditions were trans and cis double bonds, terminal vinyl end groups, saturated end groups, hydroperoxide groups and carbonyl groups. In addition to NMR measurements, molecular weight changes were monitored through intrinsic viscosity, gel permeation chromatography and low angle laser light scattering measurements. It was possible through the appropriate experimental conditions both before and during irradiation, to convert a linear, high density polyethylene to a medium density, exclusively long chain branched polymer, which was free of gel. Attempts to detect the H-link in polyethylenes irradiated to doses short of the gel dose met with little success.

The direct detection of radiation induced crosslinks in polyethylene has been a major goal of radiation chemists for many years. It was recognized as early as 1967 that solution ^{13}C nuclear magnetic resonance (NMR) spectroscopy could be used to detect structures produced in polymers from ionizing radiation. Fischer and Langbein(1) reported the first direct detection of radiation induced crosslinks (H-links) in polyoxymethylene using ^{13}C NMR. Bennett et al.(2) used ^{13}C NMR to detect radiation induced crosslinks in n-alkanes irradiated in vacuum in the molten state. Bovey et al.(3) used this technique to identify both radiation induced H-links and long chain branches (Y-links) in n-alkanes

0097-6156/84/0247-0245$06.75/0

irradiated in the molten state. Bovey *et al.* also determined that few, if any, H-links and long chain Y branches form in n-alkanes irradiated in the solid state.

Previous investigators have had little success in obtaining resonances arising from the gel structure formed during irradiation of polyethylene. Partially gelled polyethylene can be swollen with an appropriate solvent to produce a "solution" which appears suitable for ^{13}C NMR liquid measurements. However, the results of this study demonstrate that only the mobile, soluble component produces observable resonances under these conditions. It is possible that NMR dipolar interactions in the gel phase are not effectively averaged through the available polymer molecular motions. Under such circumstances, the resonances would become so broad as not to be observed in normal high resolution liquid ^{13}C NMR spectra.

The problems involved in observing resonances from gelled polyethylene were also encountered in this study. We decided, therefore, to examine polyethylene samples irradiated with absorbed doses less than the gel dose. By so doing, some of the radiation induced structural changes produced prior to gelation could be detected.

It may be instructive at this point to review some of the advantages provided by the ^{13}C NMR technique for examining soluble irradiated polyethylenes:

(a) Measurements at 50 MHz and 398 K at solution concentrations between 10 and 20 percent by weight can yield high resolution spectra displaying a sensitivity of approximately 0.5 structural units per 10,000 total carbon atoms.

(b) Each separate polyethylene structural entity gives rise to a unique array of resonances; the number and relative intensities depend upon the number and types of carbon atoms associated with each particular structural unit. Thus a number of different resonances will be associated with a structural unit. Model compounds and model polymers can lead to unequivocal assignments for the various types of polyethylene carbons.

(c) Each resonance intensity is directly proportional to the number of contributing carbon atoms. Thus there is only one proportionality constant for all resonances and there is no need to determine various extinction coefficients as required for infrared and ultraviolet spectral measurements.

(d) Saturated end groups and both short and long chain branches can be monitored independently using the ^{13}C NMR technique.

(e) Internal cis and trans double bonds and terminal vinyl groups can be monitored directly and independently with the ^{13}C NMR method.

Advances in high resolution ^{13}C NMR equipment, such as higher magnetic field strengths, improved probes, improved dynamic ranges and sophisticated dedicated computer systems have greatly increased the sensitivity of ^{13}C NMR measurements. New instruments, operating at frequencies of either 300 or 500 MHz, have extended the capability to detect long chain branching in polyethylene to as few as 1-5 branches per 100,000 carbon atoms.

The concentrations of structures produced in irradiated polyethylene are on the order of 1 per 10,000 carbon atoms for absorbed doses of approximately 2.0 Mrad.   Although the approach of examining polyethylenes irradiated with absorbed doses less than the gel dose placed a premium on sensitivity, we were able to detect the first direct radiation induced long chain branches in high density polyethylene (4).

Experimental

Irradiation.   All irradiations were performed at the University of Maryland with a 25,000 curie cobalt 60 gamma source.   The absorbed dose rate of 1.2 Mrad per hr was determined by ferrous sulfate dosimetry.   All samples were irradiated under secondary electron equilibrium conditions. Following irradiation, all samples irradiated in vacuum were annealed in vacuum at 380 K for 24 hours;   this treatment reduces the long-lived radical concentration to undetectable levels.

Materials.   The n-hexatriacontane (HTC, n-C_{36}, M.P. 75°C) samples were obtained from Aldrich Chemical Company.   Irradiations were performed in vacuum in both the solid and molten states to reproduce the results reported by both Bennett *et al.*(2) and Bovey *et al.*(3) where the H-link was reported to form.   The H-link chemical shift data from the model alkane irradiations (at 100 Mrad) could be used to identify the H-link after corresponding NMR measurements on irradiated polyethylenes.

Phillips Marlex 6003, a high density polyethylene (M_w = 53,000; M_n = 18,000) possessing predominantly saturated end groups was irradiated to 4.0 Mrad in vacuum in pellet form as supplied.   NBS 1475 is a commercial polyethylene containing 111 ppm of Irganox 1010.   Other experiments involved heating in vacuum to 500 K for 24 h prior to irradiation to 3.0 Mrad at 500 K.   The NBS 1475 sample required a greater amount of irradiation to reach the gel dose than did Marlex 6003.

Gel Fraction Determinations.   The irradiated polyethylene samples were measured for gel content by extraction with boiling xylene for 72 h.   Only one sample, NBS 1475 irradiated to 8.0 Mrad in vacuum at 300 K, showed a measurable gel content.

^{13}C NMR Measurements.   A Varian XL-200 NMR spectrometer, located at the Phillips Research Center, was used to make all of the ^{13}C NMR measurements at 50.3 MHz.   The samples were dissolved in 1,2,4-trichlorobenzene at 15 weight percent and maintained under a nitrogen atmosphere at 398 K in the probe cavity during data acquisition.   Other conditions were as follows:

Pulse angle:	90°
Pulse delay:	15.0 seconds
Acquisition time:	1.0 seconds
Spectral width	8000 Hz
Data points/spectrum:	32,000
Free Induction Decays (FID's) accumulated:	5,000 - 20,000

Double precision arithmetic was employed during data acquisition and the software included a floating point Fourier transform capability.

A factor controlling the ultimate sensitivity available is the linewidths at one-half height of the polyethylene resonances. It is desirable to create conditions which lead to the most narrow linewidths possible when examining the irradiated polyethylenes by ^{13}C NMR. At concentrations of 15 weight percent and at temperatures of 398 K, linewidths of 1.0 - 1.5 Hz were obtained at one-half height for the major polyethylene resonances for the recurring equivalent methylene units, $\delta^+\delta^+$, at 29.98 ppm from TMS as shown in Table I. Substantially larger linewidths at one-half height were obtained for solution concentrations higher than 15 weight percent. Narrow resonances lead to greater peak heights, which in turn lead to greater sensitivity.

Another problem encountered during quantitative measurements was whether peak heights could be reliably used for intensities of weak resonances. The end group resonances were of sufficient strength to allow a comparison of peak heights to integrated areas. It was clear from the narrow observed line widths (<1.0 Hz @ $\frac{1}{2}$ height) of the end group resonances that peak heights would lead to low results when used to measure molecular weight. Area measurements by spectral integration led to number average molecular weights of 18,900 and 17,800 for Marlex 6003 and NBS 1475, respectively, in good agreement with the values obtained by GPC. Number average molecular weights of approximately 12,000 were obtained for both samples when relative peak heights were used from the same spectral data. The most narrow linewidths consistently occurred for carbon resonances from chain ends and ends of branches. Consequently, relative peak heights are more reliable for weak resonances associated with structural moieties from interior sequences of the polymer chain. Measurements of the Y branch, cis and trans double bonds, carbonyl and hydroperoxide group concentrations in a range of 1-5 per 10,000 carbons were made by a comparison of peak heights because these resonances were generally too weak for reliable integral measurements.

Weighting functions can be employed to smooth the accumulated free induction decay signal to improve the signal to noise ratio upon Fourier transformation of the data. Such smoothing can lead to improved peak height ratios because it tends to reduce differences among resonance linewidths as resolution is lost with the reduction in noise level. Essentially no differences were obtained in the quantitative results for long chain branching, cis and trans double bonds, carbonyl and hydroperoxide groups based on peak heights for the weighted versus unweighted data. Results for the end groups based on peak heights were improved, however, when a smoothing function was used to increase signal to noise. This result would tend to validate the use of peak heights for those weak resonances from interior chain structural moieties. Unfortunately, the available sensitivity was pushed to the limit in this study and no recourse other than the use of peak heights appeared feasible for intensity measurements of weak resonances. The data in Tables II through VI were derived from unweighted data.

Table I

Linewidths at One-Half Height for the $\delta^+\delta^+$ Resonance at 29.98 ppm as a Function of Radiation Dose.

Sample	Radiation Dose	LineWidth at $\frac{1}{2}$ Height	FID's Completed
NBS 1475 (Fig. 9)	None	1.2 Hz	9,699
NBS 1475	2.0 Mrad at 398 K	1.2	4,482
NBS 1475 (Fig.10)	4.0 Mrad at 398 K	1.3	14,500
NBS 1475 (Fig.12)	8.0 Mrad at 398 K*	1.1	5,104
NBS 1475 (Fig.11)	3.0 Mrad (Melt, 500 K) (after 24 h. at 500 K)	1.4	7,271
Marlex 6003 (Fig. 4)	None	1.3 Hz	20,683
Marlex 6003 (Fig. 5)	2.0 Mrad at 398 K	1.6	9,617
Marlex 6003 (Fig. 8)	4.0 Mrad at 398 K	1.0	5,250
Marlex 6003 (Fig. 6)	None (after 24 h at 550 K)	1.3	4,314
Marlex 6003 (Fig.7)	1.0 Mrad at 550 K (after 24 h at 550 K)	1.2	9,302

*Sample partially gelled; soluble

Table II

^{13}C NMR Chemical Shifts Associated with the H-link

Carbon	Chemical Shifts wrt TMS, ppm		
	This Study	Bennett *et al.*(2)	Bovey *et al.*(3)
Methine	41.01	39.49	40.5
α CH$_2$	30.47	30.70	31.9
β CH$_2$	28.62	28.22	28.6
γ CH$_2$	30.60	30.19	----

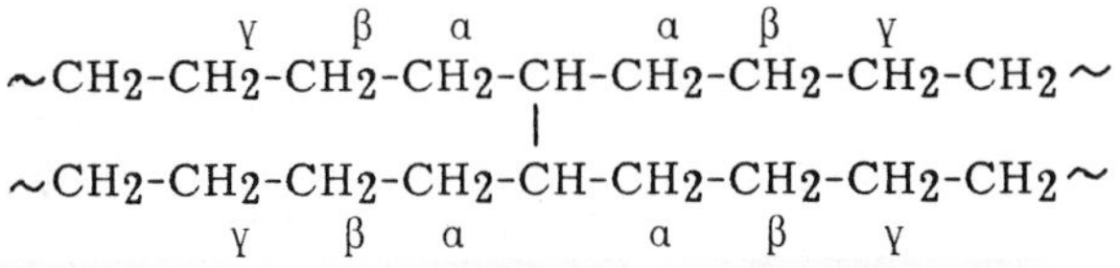

$$\begin{array}{ccccccc} \gamma & \beta & \alpha & & \alpha & \beta & \gamma \\ \sim CH_2\text{-}CH_2\text{-}CH_2\text{-}CH_2\text{-}CH\text{-}CH_2\text{-}CH_2\text{-}CH_2\text{-}CH_2 \sim \\ & & & | \\ \sim CH_2\text{-}CH_2\text{-}CH_2\text{-}CH_2\text{-}CH\text{-}CH_2\text{-}CH_2\text{-}CH_2\text{-}CH_2 \sim \\ \gamma & \beta & \alpha & & \alpha & \beta & \gamma \end{array}$$

Table III
Concentrations Following Irradiation of n-Hexatriacontane (HTC)

Structural Unit	Number of Units per 10,000 Carbon Atoms	
	Irradiated 100 Mrad in Vacuum at 298 K	Irradiated 200 Mrad in Vacuum at 355 K
Saturated End Groups[a]	562	558
Trans Double Bonds	8.6	4.9
Cis Double Bonds	Not Detected	1.5
Long Chain Branches (Y)	Not Detected	5.4
H-Links	Trace	3.3

[a]By integrated area measurements; all other measurements from relative peak heights. Calculated HTC saturated end group concentration prior to irradiation is 556 per 10,000 carbon atoms.

Table IV
Structural Concentrations Following
Irradiation of Marlex

Structural Unit	Solid State 298 K Number of Units per 10,000 Carbon Atoms		
	Before Irradiation	Irradiated 2.0 Mrad in Vacuum	Irradiated 4.0 Mrad in Air
Saturated End Groups[a]	7.3	8.6	10.1
Vinyl End Groups[a]	7.8	3.9	4.1
Long Chain Branches (Y)	1.2	2.2	N.D.
Trans Double Bonds	1.7	3.1	3.1
Cis Double Bonds	1.7	2.5	3.1
Hydroperoxide Groups	1.8	2.8	4.6
Carbonyl Groups	0.7	N.D.	1.7
$M_W \times 10^{-3}$	140[b]	164[c]	58.3[c]
$M_n \times 10^{-3}$	20	---	---
M_W/M_n	7.0	≥ 7	≥ 4

a Concentrations determined by integrated area measurements.
  Remaining concentrations determined by peak height measurements.

b Measured by GPC

c Measured by LALLS

Table V
Structural Concentrations Following Thermal Degradation
and Subsequent Irradiation of Marlex 6003 Polyethylene

Structural Unit	Melt State 550K Number of Units per 10,000 Carbon Atoms	
	Heated 550 K 24 h in Vacuum	Heated 550 K 24 h in Vacuum 1.0 Mrad @ 550 K
Saturated End Groups[a]	18.6	18.6
Vinyl End Groups[a]	14.8	9.4
Long-Chain Branches (Y)	1.9	3.6
Butyl Branches	2.3	3.1
Trans Double Bonds	1.6	1.9
Cis Double Bonds	2.0	1.8
Hydroperoxide Groups	1.9	1.9
Carbonyl Groups	N.D.	N.D.
M_w x 10^{-3}	166[b]	192[b]
M_n x 10^{-3}	13.1	14.8
M_w/M_n	12.4	12.9

a Concentrations determined by integrated area measurements.
   Remaining concentrations determined by peak height measurements.
b Measured by GPC

Table VI
Structural Concentrations Following Irradiation of NBS 1475

Structural Unit	Number of Units per 10,000 Carbon Atoms				
	Before Irradiation	2.Mrad 298 K In Vac.	4.Mrad 298 K In Vac.	8.Mrad 298 K In Vac.	3.Mrad[b] 500 K In Vac.
Saturated End Groups [c]	10.4	13.0	12.9	15.1	94.3
Vinyl End Groups[c]	5.3	2.0	2.8	---	16.1
Long Chain Branches (Y)	0.7	0.9	1.0	1.3	43.5[c]
Cis Double Bonds	1.4	3.6	1.6	2.9	1.9
Trans Double Bonds	1.4	2.7	1.5	5.1	2.6
Ethyl Branches	2.5	3.2	4.0	4.3	4.1
Butyl Branches	N.D.	N.D.	N.D.	N.D.	4.4
Hydroperoxide Groups	1.8	2.7	1.2	4.6	2.5
Carbonyl Groups	N.D.	2.2	1.3	N.D.	1.3
$M_w \times 10^{-3}$	52.8[d]	116[d]	128[d]	---	35.8[d]
$M_n \times 10^{-3}$	18.1	21.8	22.3	---	5.5
M_w/M_n	2.9	5.3	5.7	---	6.5

a Sample partially gelled; observed soluble component only.
b Sample heated in vacuum @ 500 K for approximately 24 hours
   prior to irradiation.
c Concentrations determined by integrated area measurements;
   other concentrations determined by peak height measurements.
d Measured by GPC.

The nomenclature employed to designate the various carbon atoms in the structural moieties monitored in this study is shown in Figure 1. The quantitative results given in Tables III through VI were obtained by dividing the intensity of one carbon from a structural moiety by the total carbon intensity for the entire spectrum and multiplying by 10,000. This gives the result in terms of structural units per 10,000 carbon atoms. For example, the total carbon intensity (TCI) for a ^{13}C NMR spectrum of high density polyethylene given by

$$TCI = \delta^+\delta^+ + 3(s+a) \tag{1}$$

where "s" is the average intensity observed for the 1s, 2s and 3s carbons and "a" is the allylic carbon intensity for a terminal vinyl group. The $\delta^+\delta^+$ term dominates because the repeating methylene units are by far the largest contributors to the NMR spectrum. It was set at 30,000 mm. Proportional intensities for the remaining structural entities ranged from 5 to 15 mm. Fortunately, for measurements of long chain branching, there are three α and three β carbons per Y branch; Thus the α and β carbon resonance intensities need only to be 9 mm in order to have a sensitivity of one branch per 10,000 carbon atoms.

Molecular Weight Measurements   A Waters M 150C gel permeation chromatograph (GPC) equipped with four porous silica columns: two SE 4000's, one SE 500 and one PSM 60s (available from DuPont) was used for molecular weight measurements.   A Wilkes variable wavelength high temperature infrared detector, also from DuPont, was used instead of a refractive index detector.   Low angle laser light scattering (LALLS) measurements were made with a Chromatix KMX-6 unit coupled to a DuPont Model 830 size exclusion chromatograph. The column set was the same as that employed in the Waters M 150C GPC unit. The mobile phase used in these measurements was 1,2,4-trichlorobenzene and the temperature was maintained at 403 K. Both the chromatography and light scattering measurements were made at the Phillips Research Center.

Results

The structural entities monitored as a function of absorbed dose and irradiation conditions are presented in Figure 1. The nomenclature for describing the various types of carbon atoms, as well as the observed chemical shifts (from an internal TMS standard), are also included. The chemical shifts observed for the H-link structure are shown in Table II.

Irradiation of n-Hexatriacontane (HTC).   Results from ^{13}C NMR measurements on HTC irradiated to 100 Mrad in vacuum and at room temperature (298 K) in the solid state and at 353 K in the molten state are shown in Table III. The spectra from which these data were obtained

are shown in Figures 2 and 3.   The irradiated HTC was dissolved in perdeuterobenzene to form a saturated solution; it was still possible to

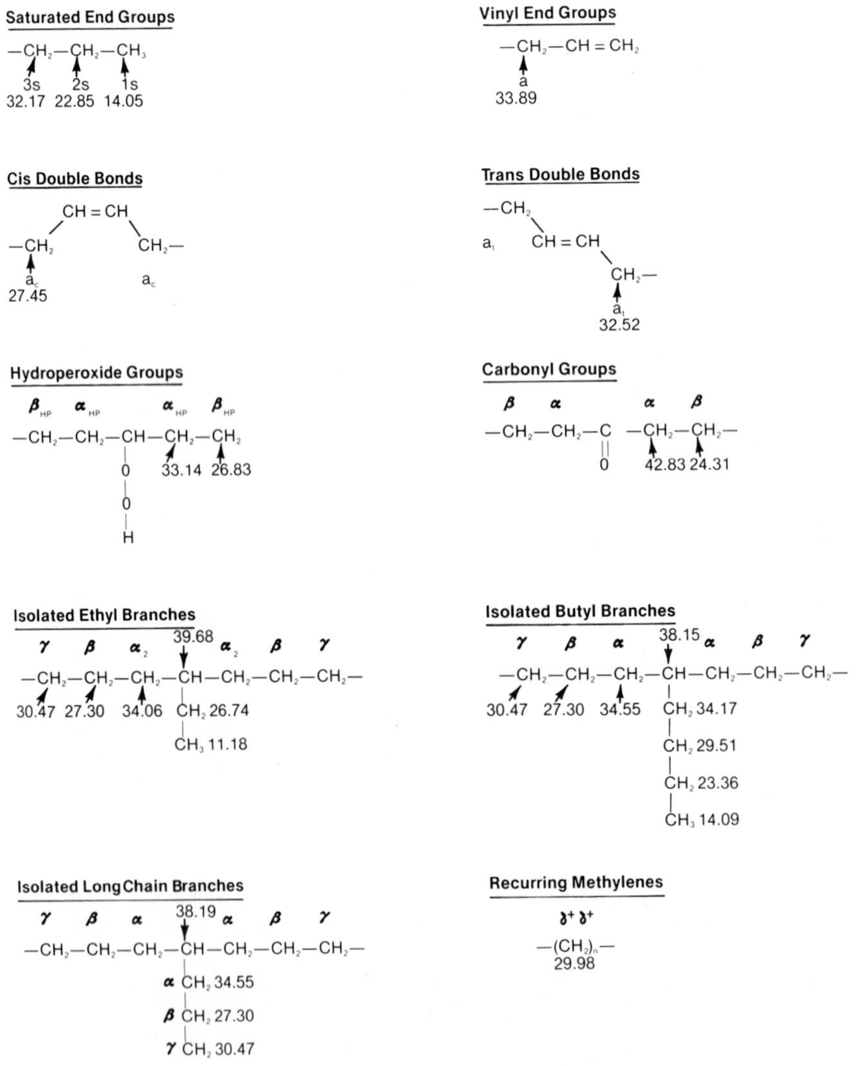

Saturated End Groups

$-CH_2-CH_2-CH_3$
3s   2s   1s
32.17 22.85 14.05

Vinyl End Groups

$-CH_2-CH=CH_2$
a
33.89

Cis Double Bonds

$CH=CH$
$-CH_2$ CH_2-
a$_t$ a$_c$
27.45

Trans Double Bonds

$-CH_2$
a$_t$ $CH=CH$
CH_2-
a$_t$
32.52

Hydroperoxide Groups

β$_{HP}$ α$_{HP}$ α$_{HP}$ β$_{HP}$
$-CH_2-CH_2-CH-CH_2-CH_2$
     O   33.14 26.83
     |
     O
     |
     H

Carbonyl Groups

β   α   α   β
$-CH_2-CH_2-C$ $-CH_2-CH_2-$
          || 42.83 24.31
          O

Isolated Ethyl Branches

γ   β   α$_2$ 39.68 α$_2$ β   γ
$-CH_2-CH_2-CH_2-CH-CH_2-CH_2-CH_2-$
30.47 27.30 34.06 CH$_2$ 26.74
                   CH$_3$ 11.18

Isolated Butyl Branches

γ   β   α   38.15 α   β   γ
$-CH_2-CH_2-CH_2-CH-CH_2-CH_2-CH_2-$
30.47 27.30 34.55 CH$_2$ 34.17
                   CH$_2$ 29.51
                   CH$_2$ 23.36
                   CH$_3$ 14.09

Isolated Long Chain Branches

γ   β   α   38.19 α   β   γ
$-CH_2-CH_2-CH_2-CH-CH_2-CH_2-CH_2-$
          α CH$_2$ 34.55
          β CH$_2$ 27.30
          γ CH$_2$ 30.47

Recurring Methylenes

δ$^+$ δ$^+$
$-(CH_2)_n-$
29.98

Figure 1. Carbon-13 NMR Chemical Shifts and Nomenclature of Structural Entities Found in Polyethylene.

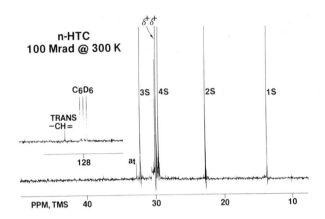

Figure 2.   Carbon-13 NMR Spectrum of n-Hexatriacontaine (HTC) Irradiated in the Solid State.

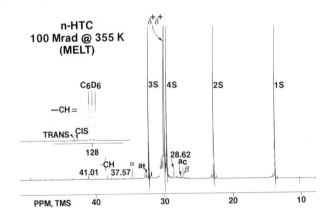

Figure 3.   Carbon-13 NMR Spectrum of n-Hexatriacontane (HTC) Irradiated in the Molten State.

obtain the NMR data at 398 K, the temperature of the polyethylene measurements.

Bennett *et al.*(2) identified the H-link in n-alkanes irradiated in vacuum in the molten state in an initial ^{13}C NMR study and reported the first NMR assignments. A corresponding H-link spectral pattern was identified in the present study after irradiation of HTC in vacuum and in the molten state. A methine resonance, confirmed by the presence of a doublet upon off-resonance decoupling, was observed at 41.01 ppm at 398 K. A methylene resonance from carbons β to the methine carbons of the H-link, was observed at 28.62 ppm. The related α and γ methylene carbon resonances were noted at 30.47 and 30.60 ppm, respectively, as indicated by Bennett *et al* for n-alkanes. A somewhat different situation is encountered in polyethylene because the strong backbone methylene resonance at 29.98 ppm and lines associated with Y branches and end groups tend to obscure a positive identification of the α and γ methylene resonances from the H-link. The overall observed resonance pattern for the HTC H-link is similar to that reported by Bennett for lower molecular weight n-alkanes although the methine chemical shift was approximately 2 ppm to lower field. By lowering the temperature to 323 K and preparing the sample in a mixed solvent of perdeuterochloroform and 1,2,4-trichlorobenzene, we were able to observe the H-link methine resonance shift from 41.01 to 39.95 ppm, which is closer to the 39.5 ppm value reported by Bennett *et al* for room temperature measurements in perdeuterochloroform. Thus under the conditions selected for a study of polyethylenes, the H-link methine resonance should be expected to reside near 41 ppm. The β methylene resonance should be visible near 28.6 ppm although the α and γ resonances would likely be totally obscured by the very strong $\delta^+\delta^+$ resonance at 29.98 ppm. Note also that the minimum detectable concentration for H-links should be higher than that for Y branches because the former has four α, β, γ carbons per structural unit, whereas the latter has three α, β, γ carbons per structural unit.

Bovey *et al.*(3) also observed both H-links and Y branches in an NMR study of a normal alkane irradiated in vacuum in the molten state. The long chain Y branch was easy to recognize in the present study because of the presence of a methine resonance at 38.19 ppm, an α methylene resonance at 34.55 ppm and a β methylene resonance at 27.30 ppm. These assignments are in excellent agreement with those of Bovey *et al.* (3) and correspond closely to the methine and α, β methylene resonances observed for an ethylene-1-octene copolymer as shown below:

Carbon	Chemical Shifts Y Branch (3)	Backbone Chemical Shifts Ethylene-1-Octene Copolymer
Methine	38.19 ppm (TMS)	38.16 ppm (TMS)
α Methylene	34.55	34.53
β Methylene	27.30	27.27

The results shown in Table III demonstrate that when HTC is irradiated in the molten state both H-links and Y branches as well as cis and trans double bonds are formed. In contrast, irradiation at room temperature produces principally trans double bonds. The linked structures formed during melt irradiation are present in reasonably comparable amounts. The number of saturated end groups, as determined by spectral integration, was not significantly affected by irradiations to 100 Mrad in either the molten or solid state. Finally, it should be pointed out that the radiation yields for structures produced by a 100 Mrad irradiation of HTC are quite low. The G values for the most abundant radiation products are less than one.

Irradiation of Marlex 6003 Polyethylene. Results of ^{13}C NMR measurements on Phillips Marlex 6003 polyethylene both prior to and just following a 2.0 Mrad irradiation in vacuum and 298 K are presented in Table IV. The spectra from which these data were obtained are shown in Figures 4 and 5. These results indicate that irradiation of high density polyethylene in vacuum in the solid state reduces the concentration of terminal vinyl unsaturations and increases the concentrations of long chain Y branches, saturated end groups and trans double bonds. The H-link could not be detected following an irradiation of Marlex 6003 in the solid state.

The ^{13}C NMR spectrum of thermally degraded Marlex 6003 is shown in Figure 6. The sample was heated in vacuum at 550 K for 24 hours. It is clear that this treatment produced additional terminal vinyl unsaturations, saturated end groups and both short and long chain branches. A subsequent irradiation to only 1.0 Mrad in vacuum at 550 K resulted in a reduction in terminal vinyl unsaturation and an increase in the number of Y branches. No resonances from an H-link could be detected. Judging from the yield of H-links in the model HTC irradiations, we should not expect to detect the presence of H-links because the level should be below 0.5 per 10,000 carbons. The copious yield of Y branches formed after irradiation may be related to the substantial quantity of terminal vinyl unsaturation produced by thermal degradation. Complete quantitative data is given in Table V. It is interesting that irradiations more extensive than 1.0 Mrad at 550K produced partially gelled polyethylene samples. A spectrum of non-gelled polyethylene produced by the combination of thermal degradation and 1.0 Mrad irradiation is shown in Figure 7.

Results of measurements on Marlex 6003 irradiated to 4.0 Mrad in air are also included in Table IV. The spectrum from which these data were obtained is shown in Figure 8. These results indicate that the radiolytic oxidation of solid high density polyethylene produces an increase in the concentration of hydroperoxide and carbonyl groups with an accompanying drastic reduction in both the number average and weight average molecular weights of the polymer.

Irradiation of NBS 1475 Polyethylene. Results of ^{13}C NMR measurements on NBS 1475 both prior to and following 2.0 and 4.0 Mrad irradiations in vacuum at room temperature are shown in Table VI. The spectra which were used to determine the concentrations listed in Table VI are shown in

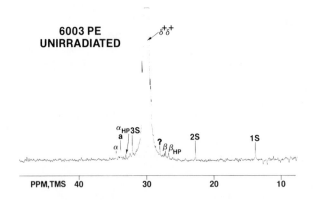

Figure 4. Carbon-13 NMR Spectrum of Phillips Marlex 6003 Prior to Irradiation.

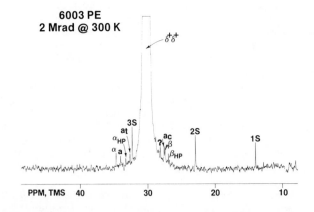

Figure 5. Carbon-13 NMR Spectrum of Phillips Marlex 6003 Irradiated in the Solid State.

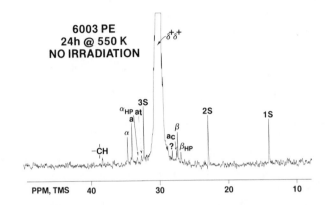

Figure 6.   Carbon-13 NMR Spectrum of Phillips Marlex 6003 Held at 550 K in Vacuum for 24 Hours.

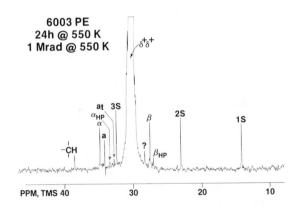

Figure 7.   Carbon-13 NMR Spectrum of Phillips Marlex 6003 Held at 550 K in Vacuum for 24 Hours and Irradiated at 550 K.

Figures 9 and 10. The results for NBS 1475 are similar to those obtained from the study of Marlex 6003 after irradiation in the solid state. Once again, the concentrations of long chain Y branches, saturated end groups and trans double bonds apparently increase with irradiation while the concentrations of terminal vinyl groups decrease with irradiation. There is some scatter in the data as might be expected for concentrations measured in the range of 1-3 per 10,000 carbon atoms. As was the case for Marlex 6003, the presence of any appreciable quantity of H-links was not detected. Thus the H-link concentration must be well below one per 10,000 carbon atoms. A preference for Y branch formation has been consistently observed in the studies of irradiated Marlex 6003 and NBS 1475 polyethylenes.

NBS 1475 polyethylene was also subjected to thermal degradation for 24 hours at 500 K and then irradiated in vacuum also at 500 K to 3.0 Mrad. The quantitative results are given in Table IV and the ^{13}C NMR spectrum which yielded these data is shown in Figure 11. As remarkably indicated in Figure 11, the formerly linear NBS 1475 is now extensively long chain branched. From a comparison of relative peak areas, the concentration of long chain branches is now approximately 44 per 10,000 carbon atoms. There are two more products formed under these conditions: the concentration of saturated end groups is now 94 per 10,000 carbon atoms and the concentration of terminal vinyl groups is 16 per 10,000 carbon atoms. The combined thermal and irradiation treatments converted a high density polyethylene (0.978) to a medium density (0.947) polyethylene without introducing short chain branches or dramatically increasing the molecular weight.

The molecular weight measurements made on this sample show that both the number average and weight average molecular weights decreased following the combined thermal treatment and irradiation. Extremely weak resonances were perceptible near 41.1 ppm and 28.6 ppm, which suggests that H-links were also formed during the high temperature melt irradiation. The H-link concentration is estimated at approximately 1 per 10,000 carbon atoms.

Results of measurements on a sample of NBS 1475 irradiated to 8.0 Mrad in vacuum at 300 K are also shown in Table VI. The spectrum from which these data were obtained is shown in Figure 12. This experiment was the only one performed in this study where the gel dose was exceeded. An attempt was made to dissolve the entire sample in 1,2,4-trichlorobenzene but only a clear swollen gel was the result. The subsequent ^{13}C NMR spectrum indicated that only the mobile, soluble component produced observable resonances. The only significant structural unit identified in the soluble component was trans double bonds; a trace amount of long chain Y branches was also observed, but H-links could not be detected in the soluble fraction.

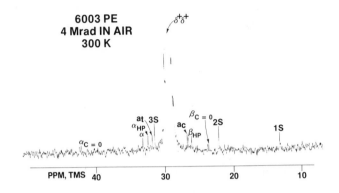

Figure 8. Carbon-13 NMR Spectrum of Phillips Marlex 6003 Irradiated in Air in the Solid State.

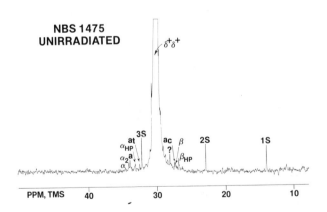

Figure 9. Carbon-13 NMR Spectrum of NBS 1475 Prior to Irradiation.

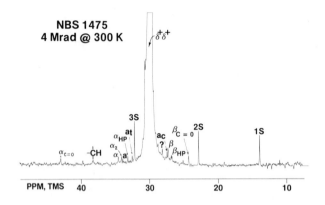

Figure 10. Carbon-13 NMR Spectrum of NBS 1475 Irradiated in the Solid State.

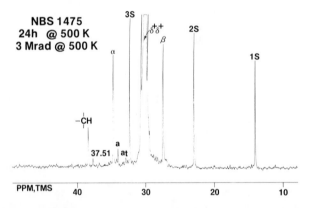

Figure 11. Carbon-13 NMR Spectrum of NBS 1475 Held at 500K for 24 Hours and Irradiated at 500K.

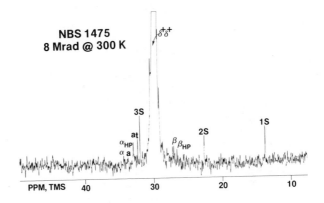

Figure 12. Carbon-13 NMR Spectrum of NBS 1475 Irradiated to 8 Mrad in the Solid State (Soluble Portion Only).

Discussion

Radiation induced structural changes in polyethylene have been investigated for over 30 years. These studies have shown that the principal chemical changes which occur during irradiation are:

(1) production of molecular hydrogen,
(2) production of trans double bonds,
(3) disappearance of terminal vinyl unsaturations and
(4) production of intermolecular links.

Although the production of molecular hydrogen was not monitored in this study, ^{13}C NMR was an excellent technique to follow the production of internal double bonds, intermolecular links and the disappearance of terminal vinyl groups. Additionally, it was observed that saturated end groups are apparently produced during irradiation of polyethylene.

The production of trans double bonds during irradiation of polyethylene has been studied by many investigators principally through infrared spectroscopy. Although cis double bonds were probably considered as a product of irradiation, conventional infrared methods could not be used to monitor the formation of cis double bonds. The ^{13}C NMR results reported in this study show that both cis and trans double bonds are produced during irradiation of molten polyethylene while trans double bond formation predominates during irradiation of solid polyethylene. In the irradiated crystalline HTC and in the soluble portion of a partially gelled NBS 1475 polyethylene, which had been irradiated to 8.0 Mrad, trans double bonds were the major product formed. These results suggest that trans double bonds are a major product produced upon irradiation in the crystalline regions of polyethylene while both cis and trans double bonds are formed in amorphous regions.

The disappearance of terminal vinyl groups during irradiation of polyethylene has also been studied by many investigators who utilized infrared methods. The ^{13}C NMR results reported in this study are similar to the infrared results reported by Okada and Mandelkern (6). Lyons (7,8) and Mandelkern (6) have proposed that vinyl disappearance is related to the formation of radiation induced links in polyethylene. Mandelkern has also proposed that vinyl disappearance is related to a decrease in the production of molecular hydrogen during irradiation (6).

One of the major new structural units identified in the present study of irradiated polyethylenes is the long chain Y branch, which forms during irradiation of polyethylenes in vacuum both in the solid state and in the high temperature molten state. Long chain Y branches are also formed in HTC irradiated in vacuum in the molten state, as previously noted by Bovey et al. (3) after irradiation of n-alkanes in the molten state. The NMR results shown in Tables IV, V and VI demonstrate that the terminal vinyl group concentration decreases as the long chain Y branch concentration increases. Upon comparing the results from the irradiated NBS 1475 to those obtained from irradiated Marlex 6003, it was noted that the yield of long chain Y branches was greater for Marlex 6003, which had a greater initial terminal vinyl concentration. Although some caution should be exercised because NBS 1475 contained an antioxidant whereas Marlex 6003 did not, these observations do suggest that some of the terminal vinyl groups may be reacting with secondary alkyl radicals to form long chain Y branch radicals. Such a hypothesis was first advanced by B. J. Lyons[7] as shown below:

$$\begin{matrix} \sim CH_2 \\ \quad \diagdown \\ \quad\quad CH\cdot \\ \quad \diagup \\ \sim CH_2 \end{matrix} + \ CH_2=CH-CH_2 \sim \longrightarrow \begin{matrix} \sim CH_2 \diagdown \\ \quad\quad\quad CH-CH_2-CH-CH_2 \sim \\ \sim CH_2 \diagup \end{matrix} \quad (2)$$

The subsequent reactions of the Y branch radical are unknown at this time; although chain transfer to form a long chain Y branch appears to be a likely possibility. Thus Y branch formation would result from a propagating step in the free radical chemistry, as opposed to a terminating step which is required to produce H-links. There is also no explanation for the observation that the disappearance of terminal vinyl groups exceeds the formation of long chain Y branches. Other means of vinyl disappearance and other linking reactions are possible. The establishment of a satisfactory mechanism accounting for all the observed products must await further experimental evidence. Strong support for Lyons' hypothesis is provided by the experimental results of Mandelkern *et al.*(6), which show that the gel dose for hydrogenated polyethylene is approximately three times that of the same polyethylene containing one terminal vinyl group per molecule. The results of the present study, which produced the first direct detection of radiation induced long chain branching in polyethylene, also supports Lyons' hypothesis.

The most important finding in this study is that the yield of long chain Y branches appears to be much greater than the yield of H-links when polyethylene is irradiated in vacuum in the solid state with absorbed doses less than the gel dose. The H-link could possibly be detected in only one polyethylene sample and that was after NBS 1475 was heated to 500 K for 24 hours in vacuum and subsequently irradiated in vacuum to 3.0 Mrad at 500 K. There are two possible explanations as to why the H-link went unobserved in the ^{13}C NMR spectra of the other irradiated polyethylenes: (1) the H-link could have remained below the detection limit of approximately 0.5 units per 10,000 carbon atoms or   (2) the H-links possessed so little molecular mobility as to go unobserved because of either dipolar broadening or relaxation effects. The latter explanation is not entirely plausible because the polyethylene matrix surrounding the H-link is sufficiently mobile to permit an observation of other structural species. Also the existence of persistent dipolar interactions within an H-link would require it to be substantially less mobile than a long chain Y branch. It is reasonable to conclude that the formation of long chain Y branches in polyethylene irradiated with absorbed doses less than the gel dose is significantly more important than the formation of H-links.

Perhaps the most surprising result obtained in this study was the consistent observation of an apparent small increase in saturated end group concentration following irradiation of solid polyethylenes in vacuum, with no accompanying decrease in the number average molecular weight of the polyethylenes (Tables IV and VI). This apparent increase in saturated end group concentration is not well understood at this time but could arise from saturation of terminal vinyl groups during irradiation or from migration of the terminal vinyl group to an inner position in the chain. Radiation induced scission through β cleavage could also lead to saturated end groups but appears unlikely because of the observed reduced content of terminal vinyl groups and the absence of sufficient linking products to account for the difference:

$$\sim CH_2CH_2\overset{\bullet}{C}HCH_2CH_2\sim \longrightarrow \sim CH_2CH_2CH=CH_2 \quad + \quad \cdot CH_2\sim \qquad (3)$$

This reaction is known to occur during thermal degradation of polyethylene but proof of its contribution to the products formed during irradiation of polyethylenes in the solid state will have to await further evidence.

As can be seen in Tables IV and V, high temperature thermal degradation of polyethylene in vacuum leads to the formation of terminal vinyl unsaturation, saturated end groups, butyl branches and long chain Y branches. Following thermal degradation the resulting changes in the concentrations of terminal vinyl groups and Y branches are collectively close to the change in the observed concentration of saturated end groups. This result strongly supports the occurrence of β cleavage during thermal degradation in vacuum at temperatures of 500-550 K. A subsequent irradiation in vacuum at 500-550 K leads to an increase in molecular weight, a substantial increase in long chain Y branches and a loss of terminal vinyl groups. These results demonstrate that long chain Y branches can be a product of thermal degradation and that this reaction can be enhanced by irradiation although there is no longer a material balance between the formation of Y branches and the loss of terminal vinyl groups. There must be a relationship between the initial terminal vinyl content and radiation induced formation of Y branches because the yield of radiation induced Y brances is directly proportional to the initial concentration of terminal vinyl groups in the polyethylene.

Finally, irradiation of polyethylene in the solid state in the presence of air precludes many of the products observed following irradiation in vacuum. There is a drastic reduction in molecular weight and a substantially reduced yield of long chain Y branches. There is still a loss of terminal vinyl groups and an apparent increase in the number of saturated end groups as shown in Table IV. The hydroperoxide and carbonyl content increase as expected. Overall, these observations demonstrate that radiolytic oxidation reactions are effective competing reactions to the linking reactions observed following vacuum irradiations.

Conclusions

1. Long chain Y branches are one of the principal products formed during irradiation of high density polyethylenes in vacuum both in the solid and molten states at absorbed doses below the gel dose.
2. The concentration of H-links remained below 0.5 per 10,000 carbon atoms for absorbed doses less than the gel dose.
3. Cis double bonds are formed during both ambient and high temperature irradiations of polyethylenes but appear restricted to amorphous regions of solid polyethylenes.
4. Saturated end groups are apparently produced during irradiation of polyethylene.

5. Trans double bonds appear to be a major product formed in the crystalline regions of irradiated polyethylenes.
6. Irradiation of solid polyethylene in the presence of air leads to a reduction in the concentrations of branches, an increase in the concentrations of hydroperoxide and carbonyl groups and a decrease in molecular weight.
7. Thermal degradation of polyethylene in vacuum leads to chain scission with the production of additional saturated and vinyl end groups and both short and long chain branches.
8. The formation of long chain Y branches appears to be related to the disappearance of terminal vinyl groups in irradiated polyethylene.

This study provides a method of characterization that can be usefully applied by others in studies of irradiated polyethylenes and other polymers. Use of the powerful NMR technique will undoubtedly yield further significant information about the radiation chemistry of polymers.

Acknowledgments

The authors thank Mr. J. R. Donaldson of the Phillips Petroleum Company for performing the liquid ^{13}C NMR measurements. The authors also wish to thank Mr. C. H. Leigh of the Phillips Petroleum Company for performing both the GPC and LALLS analyses.

Several stimulating discussions with Dr. D. L. Van der Hart of the U.S. National Bureau of Standards and with Dr. J. D. Hoffman of the University of Maryland also contributed to this work.

Literature Cited

1. Fischer, H.; Langbein, W. Kolloid Z 1967, 216-217, 329.
2. Bennett, R. L.; Keller, A.; Stejny, J.; Murray, M. J. Polym. Sci., Polym. Chem. Ed. 1976, 14, 3027.
3. Bovey, F. A.; Schilling, F. C.; Cheng, H. N. in "Advances in Chemistry Series No. 169"; Allara, D. L.; Hawkins, W. L., Eds.; American Chemical Society: Washington, D.C., 1978; pp. 133-141.
4. Randall, J. C.; Zoepfl, F. J.; Silverman, J. Makromol, Chemie, Rapid Commun. 1983, 4, 149.
5. Hoeve, C. A. J.; Wagner, H. L.; Verdier, P. H. J. Res. Nat. Bur. Stand., Part A 1972, 76, 137.
6. Mandelkern, L. in "The Radiation Chemistry of Macromolecules," Vol. 1, Dole, M. Ed.; Academic Press: N.Y., 1972; Chapter 13.
7. Lyons, B. J. Polym. Prepr., Am. Chem. Sec., Div. Polym. Chem. 1967, 8, 672.
8. Lyons, B. J. in "International Conference on Radiation Processing for Plastics and Rubber"; The Plastics Institute: London (UK), 1981; Chapter 5.

RECEIVED October 24, 1983

INDEXES

Author Index

Subject Index

C

Production by Frances Reed
Indexing by Susan Robinson
Jacket design by Anne G. Bigler

Elements typeset by Hot Type Ltd., Washington, D.C.
Printing and binding by Maple Press Co., York, Pa.

RECENT ACS BOOKS

"Geochemical Behavior of Disposed Radioactive Waste"
Edited by G. Scott Barney, James D. Navratil, and W. W. Schulz
ACS SYMPOSIUM SERIES 246; 413 pp.; ISBN 0-8412-0827-1

"Size Exclusion Chromatography: Methodology and
Characterization of Polymers and Related Materials"
Edited by Theodore Provder
ACS SYMPOSIUM SERIES 245; 392 pp.; ISBN 0-8412-0826-3

"Industrial-Academic Interfacing"
Edited by Dennis J. Runser
ACS SYMPOSIUM SERIES 244; 176 pp.; ISBN 0-8412-0825-5

"Characterization of Highly Cross-linked Polymers"
Edited by S. S. Labana and Ray A. Dickie
ACS SYMPOSIUM SERIES 243; 324 pp.; ISBN 0-8412-0824-9

"Polymers in Electronics"
Edited by Theodore Davidson
ACS SYMPOSIUM SERIES 242; 584 pp.; ISBN 0-8412-0823-9

"Radionuclide Generators: New Systems
for Nuclear Medicine Applications"
Edited by F. F. Knapp, Jr., and Thomas A. Butler
ACS SYMPOSIUM SERIES 241; 240 pp.; ISBN 0-8412-0822-0

"Polymer Adsorption and Dispersion Stability"
Edited by E. D. Goddard and B. Vincent
ACS SYMPOSIUM SERIES 240; 477 pp.; ISBN 0-8412-0820-4

"Assessment and Management of Chemical Risks"
Edited by Joseph V. Rodricks and Robert C. Tardiff
ACS SYMPOSIUM SERIES 239; 192 pp.; ISBN 0-8412-0821-2

"Chemical and Biological Controls in Forestry"
Edited by Willa Y. Garner and John Harvey, Jr.
ACS SYMPOSIUM SERIES 238; 406 pp.; ISBN 0-8412-0818-2

"Chemical and Catalytic Reactor Modeling"
Edited by Milorad P. Dudukovic and Patrick L. Mills
ACS SYMPOSIUM SERIES 237; 240 pp.; ISBN 0-8412-0815-8

"Archaeological Chemistry--III"
Edited by Joseph B. Lambert
ADVANCES IN CHEMISTRY SERIES 205; 324 pp.; ISBN 0-8412-0767-4

"Molecular-Based Study of Fluids"
Edited by J. M. Haile and G. A. Mansoori
ADVANCES IN CHEMISTRY SERIES 204; 524 pp.; ISBN 0-8412-0720-8